I0044176

ELECTRICAL POWER

PROJECTS and FACTS

Stephen P. Tubbs, M.Sc., P.E.

Second Edition

Copyright 1996, 1997 by Stephen P. Tubbs
1344 Firwood Dr.
Pittsburgh, PA 15243

Type reset, references updated, and new cover in 2003

All rights reserved. No part of this book may be reproduced, in any form or by any means, without permission in writing from the publisher.

The author of this assumes no responsibility for the safe construction and use of equipment made using this book. It is the responsibility of the reader to use common sense and safe electrical and mechanical practices.

ISBN 0-9659446-1-1

CONTENTS

INTRODUCTION

ELECTRICAL POWER PROJECTS AND FACTS considers unusual electrical power topics of interest to practical users of electrical power and to experimenters. It is a compilation of reports that were previously sold individually.

Each *ELECTRICAL POWER PROJECTS AND FACTS* project plan and report is written in a down-to-earth and easy to understand form. Higher mathematics and complicated wiring diagrams are not used.

The project plans and reports are unique. You won't find them elsewhere.

I am a licensed electrical engineer. My background and interests are in the generation and application of electric power. These project plans and reports were written using knowledge gained through years of working in industry and education. The plans describe actual equipment I have constructed. Information that is presented comes from many sources: books, catalogs, microfilm files, experts in the field, experiments, and most of all experience.

When doing electrical projects:
1) Make good solid connections using wire with good insulation.
2) Avoid touching electrically live or rotating parts.
3) Double check every circuit before energizing it.
4) Avoid electrocution. Remember that many of the voltages we take for granted (the 115 volts used in the home, for example) are high enough for electrocution.

Your safety is your responsibility. Use common sense and safe electrical and mechanical practices.

1. SAVE MONEY GENERATING ELECTRICITY?

This contains information to evaluate the worth of electric power generation opportunities available with wind, water, and solar power.

Wind generators, water turbines, and solar electric panel (photovoltaic) systems are evaluated for their practicality in generating electricity. Information is given to help the potential user do a rough estimate of: 1) The amount of power he has available. 2) How much it would cost to install equipment to use that power.

If the reader decides to build his own electric generating system, he should investigate further before purchasing equipment or starting construction.

*** THERE IS NO FREE ENERGY

Looking at the wind blow the leaves, watching water cascade down a stream, or feeling the hot sun, one might believe that he is observing free energy. It is only free if it is left as is. When it is converted to another form there is an expense, the cost of the conversion equipment. The cost of the equipment can be reasonable or beyond reason. Each possible energy source should be evaluated for cost before buying equipment.

*** TYPES OF POWER GENERATION SYSTEMS CONSIDERED

DC (Direct Current) electricity is usually generated in small scale power systems because it is easier to handle than AC (Alternating Current). Only DC generation will be considered in this chapter.

Wind generators and water turbines are considered as drivers of DC generators. Solar electric panels (photovoltaic panels) are considered as sources of DC power.

Because of the variable nature of sources and for a greater short term power capability, all systems are assumed to be connected to "deep cycle" lead-acid battery storage banks. "Deep cycle" lead-acid batteries are the type used on electric golf carts. They can survive many more charge/discharge cycles than automobile batteries.

It is assumed that the user will be connecting equipment to the battery storage bank's DC output or to the output of an inverter. If an inverter is used its input would be connected to the battery storage bank's DC output. Then the inverter would change DC to 120 volts 60 Hz AC by electronic switching.

*** ELECTRIC POWER

Electric energy is measured in units of kiloWatthours (kWh). It is the product of electric power in kiloWatts (kW) multiplied by the time of use in hours (h). Power in Watts is converted to power in kiloWatts by dividing by 1000.

As an example, consider five 100 Watt light bulbs in continuous operation for one month. Electric power equals 5 x 100 / 1000 = .5 kW. Time equals 24 x 30.4 = 730 Hours. Electric energy usage equals .5 x 730 = 365 kWh. Presently, utilities charge $.10 to $.20 per kWh. The cost of using the five 100 Watt light bulbs for one month at $.10/kWh is 365 x .1 = $36.50.

*** WIND GENERATED ELECTRICITY

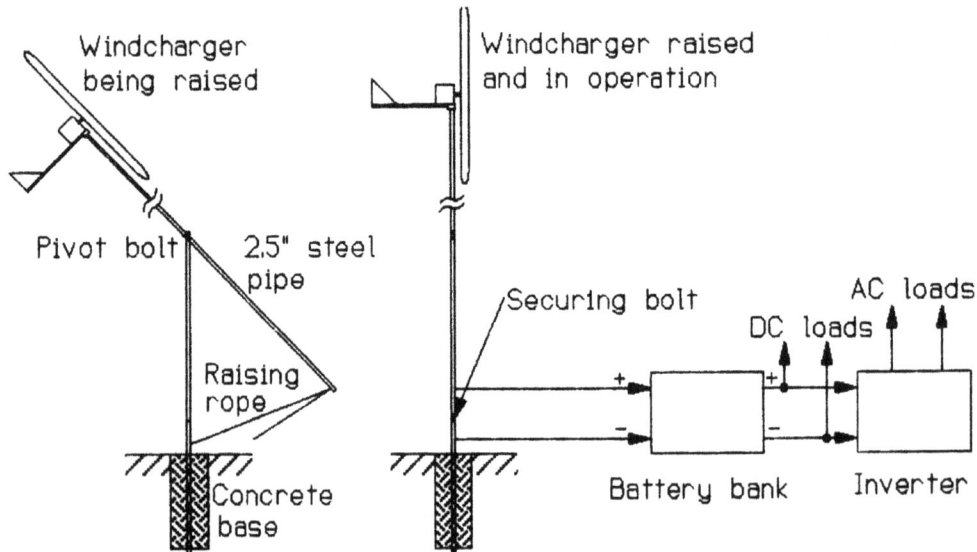

Figure 1-1 Wind generator system.

Is a wind generator a viable power source for you? Answer the following questions:

1) Do you have enough wind? Usually an average wind speed of at least 11 to 14 mph is required for a successful wind generator system. Is the wind steady? Steady winds are better suited to the operation of a wind generator than winds that vary greatly in speed.

A rough estimate of available wind speeds can be made from the following data from the U.S. Weather Bureau, 1955.

Wind Speed (mph)   Wind Effects Observed on Land

Less than 1.............. Calm; smoke rises vertically
1 to 3..................... Direction of wind shown by smoke drift; but not by wind vanes
4 to 7..................... Wind felt on face; leaves rustle; ordinary vane moved by wind
8 to 12.................... Leaves and small twigs in constant motion; wind extends light flag
13 to 18.................. Raises dust, loose paper; small branches are moved
19 to 24.................. Small trees in leaf begin to sway; crested wavelets form on inland
                                      waters
25 to 31.................. Large branches in motion; whistling heard in telegraph wires;
                                      umbrellas used with difficulty
32 to 38.................. Whole trees in motion; inconvenience felt walking against wind
39 to 46.................. Breaks twigs off trees; generally impedes progress

2) Will your wind speeds produce enough power? Maximum theoretical wind power and realistically practical available electrical power are tabulated in the following chart for different wind speeds and wind generator rotor diameters. The maximum theoretically available power is not totally convertible to electricity due to wind generator and electrical equipment inefficiencies. The practical available powers are close to those given by wind generator manufacturers. The rotor diameter in this chart refers to the outside diameter of the blades on a "classic" propeller-like wind generator.

THEORETICAL/PRACTICAL AVAILABLE POWER (WATTS)	WIND SPEED (MPH)	ROTOR DIAMETER (FT.)
51/42	10	5
170/85	10	9
440/250	10	15
440/250	20	5
1300/830	20	9
3700/2500	20	15
1700/550	30	5
5100/1000	30	9
20000/3000	30	15

3) Will zoning restrictions keep you from placing a wind generator on your property?

4) Will safety be a problem? Large wind generators are not toys and do have the potential to cause damage to people and things around them.

****

The following case study provides an estimate of the cost per kWh for energy generated by a wind generator in continuous operation.

Assumptions:
....   5 year lifetime at full continuous operation before replacement or major repair is necessary.
....   20 mph continuous winds, no hurricanes, no dead air.
....   Rotor blade diameter is 9 feet.
....   Battery bank is capable of 5 hours of 830 Watt discharge. It uses "deep cycle" batteries of the type used on electric golf carts.
....   Costs are estimates. They are from catalogs and word of mouth "thought to be reasonable" estimates. They are not to be used as more than rough estimates.

ITEM	COST
Wind generator, top of tower assembly	$1,640
One 40' 2.5" od steel pipe for tower	100
Two 20' 2.5" od steel pipes for tower legs into ground	100
Two 2.5" pipe caps and two ½" diameter, 9" long bolts with nuts	15
1.5 yds. of concrete to anchor tower legs	100
12 volt lead-acid deep cycle batteries	340
Battery voltage controller	320
400' #4 AWG cable	230
Installation cost 200 hrs. at $10/hr.	2,000
Maintenance cost 100 hrs. at $10/hr.	1,000
2500 Watt inverter	1,275

TOTAL $5,845.00 without inverter
$7,120.00 with inverter

kWh of electricity generated in 5 years = (.83 kW)x(5 yrs. x 365 days per yr. x 24 hrs. per day) = 36354 kWh
Cost per kWh without inverter = 5845/36354
 = $.16/kWh
Cost per kWh with inverter = 7120/36354
 = $.20/kWh

The speed of the wind is critical. In this example if the wind speed had been an average of 10 mph then the generated power would have been reduced to about one tenth. With 10 mph winds the cost of generated power would be from $1.6 to $2.0 per kWh.

*** WATER POWER GENERATED ELECTRICITY

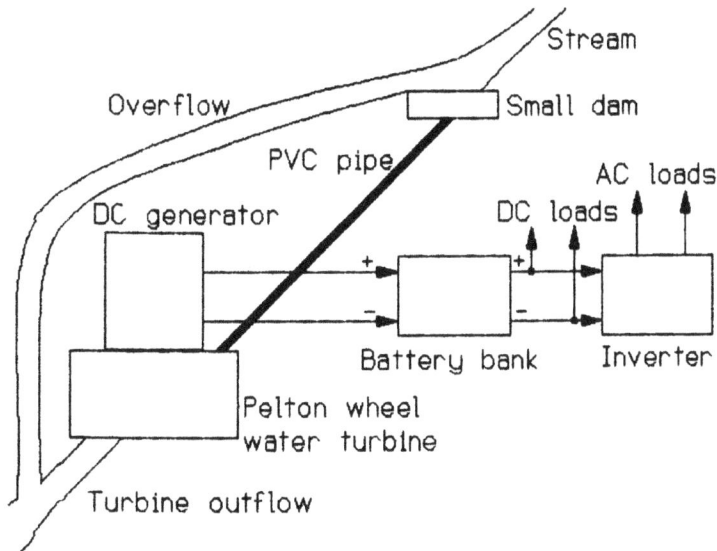

Figure 1-2 Water powered generator system.

Is a water powered generator a viable source for you? Answer the following questions:

1) What is your stream's water flow and water drop (head)?

Water flow is the amount of water going down the stream in cubic feet per second.

If the flow is low, measure it by channeling all the stream's water into a pipe or spout to a bucket. Then, see how long it takes to fill the bucket. Flow rate equals the volume of the bucket in cubic feet divided by the time for it to fill in seconds. For flows too great for a bucket, find a 30 feet length of the stream that has a roughly constant cross-sectional area. Then, time how long it takes a small piece of wood to float with the current the 30 feet. The flow rate in cubic feet per second is the cross sectional area of the stream in square feet times 30 feet divided by the passage time of the piece of wood. A third method is to use a weir dam. Weir dams are a standard method and are more accurate, but they are time consuming to build. For weir dams, see Stoner's book in the references.

The water drop or head is the distance the surface of the water will drop from the input pipe or trough of the system to the system output. Use a carpenter's level to sight from the point you are considering as the input to above the point you are considering as the output. Measure the drop in feet.

2) Can your stream generate enough power? Power capacity is determined by the water drop distance (head) and the water flow rate. The following chart shows maximum theoretical water power and practical, convertible to electricity, water power versus head and flow rate.

THEORETICAL/PRACTICAL AVAILABLE POWER (WATTS)	WATER HEAD (FT.)	FLOW RATE (CU.FT./SEC)
93/28	2	.55
370/110	2	2.2
1500/450	2	8.8
6000/1800	2	35.0
93/28	8	.14
370/110	8	.55
1500/450	8	2.2
6000/1800	8	8.8
93/28	32	.034
370/110	32	.14
1500/450	32	.55
6000/1800	32	2.20
93/28	128	.0086
370/110	128	.034
1500/450	128	.14
6000/1800	128	.55

The equation used for theoretical power is = (Head)(Flow Rate) x 84.6.

3) Will environmental restrictions keep you from damming up a stream?

4) Will safety be a problem? Don't leave exposed rotating machinery in a place that people can walk or fall into.

****

The following case study provides an estimate of the cost per kWh for energy generated by a Pelton water turbine in continuous operation.

Assumptions:
....   5 year lifetime at full continuous operation before replacement or major repair is necessary.

....   Continuous water flow at .55 cubic feet per second, 32 foot head, no floods, and no droughts.

....   Battery bank is capable of 5 hours of 830 Watt discharge. It uses "deep cycle" batteries of the type used on electric golf carts.

....   Costs are estimates. They are from catalogs and word of mouth "thought to be reasonable" estimates. They are not to be used as more than rough estimates.

ITEM	COST
Pelton water turbine including electric generator	$990
200' of 4" id PVC pipe (this assumes a 10% stream bed grade)	280
Pipe hardware	100
12 volt lead-acid deep cycle batteries	340
Battery voltage controller	320
400' #4 AWG cable	230
Installation cost 200 hrs. at $10/hr.	2,000
Maintenance cost 100 hrs. at $10/hr.	1,000
2500 Watt inverter	1,275

TOTAL $5,260.00 without inverter
$6,535.00 with inverter

kWh of electricity generated in 5 years = (.83 kW)x(5 yrs. x 365 days per yr. x 24 hrs. per day)
= 36354 kWh
Cost per kWh without inverter = 5260/36354
= $.14/kWh
Cost per kWh with inverter = 6535/36354
= $.18/kWh

*** SOLAR GENERATED ELECTRICITY

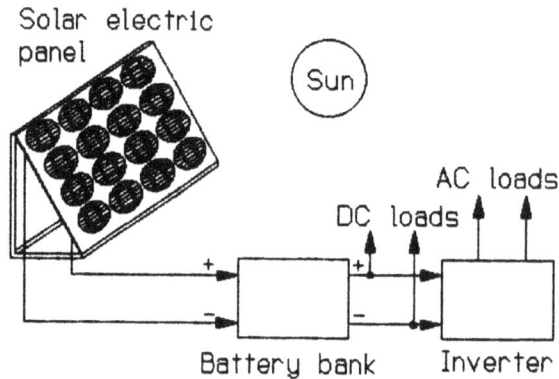

Figure 1-3 Solar electric power system.

Is solar generated electricity a viable source for you?
(Note: Do not confuse solar electric generation with solar heating. Solar heating systems do not produce electricity. Solar heating systems are economical in many regions of the country.)

Answer the following questions:

1) How many square feet of solar electric panels would be needed to generate the electric power you need? The map of the U.S. in Figure 1-4 shows the average (including day, night, and cloudy conditions) DC electrical power available from a typical 10% efficient 1 square foot solar electric panel in different U.S. regions.

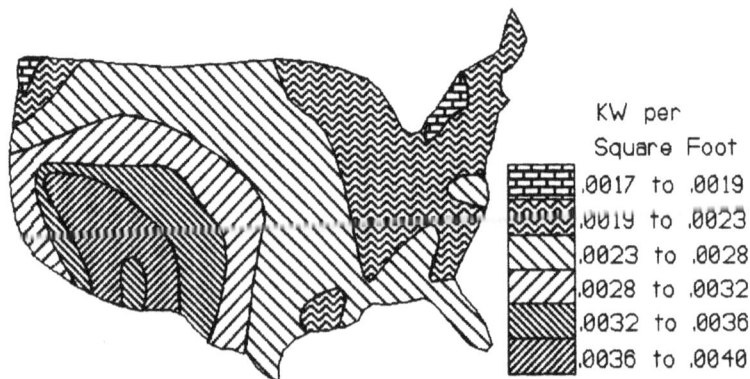

KW per
Square Foot
.0017 to .0019
.0019 to .0023
.0023 to .0028
.0028 to .0032
.0032 to .0036
.0036 to .0040

Figure 1-4 Average electric power available per square foot.

2) Solar electric panels cost about $70.00 per square foot. Do you want to spend that much?

****

The following case study provides an estimate of the cost per kWh for energy generated by a solar electric panel system.

Assumptions:

....   5 year lifetime at full continuous operation before replacement or major repair is necessary.

....   The sun shines with .04 kW per square foot on the average. This includes day, night, and cloudy conditions.

....   The sun shines 12 hours per day.

....   The efficiency of the solar electric panels is 10%.

....   The average power generated by the solar electric panels is 830 Watts.

....   Battery bank is capable of 12 hours of 830 Watt discharge. It uses "deep cycle" batteries of the type used on electric golf carts.

....   Costs are estimates. They are from catalogs and word of mouth "thought to be reasonable" estimates. They are not to be used as more than rough estimates.

ITEM	COST
Solar electric panels	$30,000
Frames to hold the solar electric panels	3,500
12 volt lead-acid deep cycle batteries	1,600
Battery voltage controller	320
400' #4 AWG cable	230
Installation cost 200 hrs. at $10/hr.	2,000
Maintenance cost 100 hrs. at $10/hr.	1,000
2500 Watt inverter	1,275

TOTAL $38,650.00 without inverter
$39,925.00 with inverter

kWh of electricity generated in 5 years = (.83 kW)x(5 yrs. x 365 days per yr. x 24 hrs. per day)
= 36354 kWh
Cost per kWh without inverter = 38650/36354
= $1.06/kWh
Cost per kWh with inverter = 39925/36354
 = $1.10/kWh

*** REFERENCES

Alternative Energy Engineering
P.O. Box 339, Redway, CA 95560
1-800-777-6609
http://www.alt-energy.com
    Sells wind, water, and solar power equipment. Provides prices and useful technical information.

Stoner, Carol H. (ed.), *Producing Your Own Power* (Emmaus, PA: Rodale Press, 1974).
    General descriptions of wind, water, and other "alternate energy" equipment. Good section on water power. Library of Congress number: TJ153 .S795.

Park, Jack, *The Wind Power Book* (Palo Alto, CA: Chesire Books, 1981).
    Excellent descriptions of windmill systems built by Jack Park and others. Library of Congress number: TK1541 .P37.

Check your library and bookstores. There have been many books written on alternate energy.

2. HIGH VOLTAGE SUPPLY CIRCUITS

*** WHAT ARE HIGH VOLTAGES?

The definition of high voltage depends on who is defining it. A designer of computer circuitry might consider 12 volts to be high voltage, since his circuitry operates at lower voltages. However, to those working in the electrical power field, high voltage is considered to start at about 600 volts. In this chapter the circuits and methods can produce thousands to millions of volts.

Power is related to voltage. There are high and low power voltage sources. Often high voltage sources are high power sources. However, it is not necessary for the voltage to be high to produce high power. An example is a car battery. A car battery produces high power while starting a car, but at low voltage. It is also possible for a high voltage source to have a low power capability. This is seen with a car's spark coil system. A car's spark coil system produces low power for the spark plugs, but at a high voltage.

One thing common to high voltage sources is the danger of electrocution. All high voltage sources should be treated as potentially very dangerous.

*** TRANSFORMER

A transformer transforms or changes AC (Alternating Current) electricity from one voltage level to another by electromagnetic induction. AC voltage input to a transformer is converted to alternating magnetism in the transformer's core, then that is converted back to AC voltage at the transformer's output. Depending on the transformer's design, the voltage level of the output may be greater than, lesser than or equal to the voltage level that was input.

Transformers will not operate with continuous DC (Direct Current) voltage.

Often, there are more than two windings on a transformer. Here only two winding transformers will be considered. The operation of transformers with more than two windings is similar.

Figure 2-1 shows a two winding laminated steel E type transformer. It is called an E type because the lower part of the core looks like an E when the upper part of the core is removed. There are two windings on the transformer. Each winding consists of turns of magnet wire around the core material. Magnet wire is copper wire with a thin coating of insulating varnish. The primary winding receives input voltage. Input voltage on the primary winding causes AC currents to flow through it. Those currents produce alternating (changing with time) magnetic fields in the core. The alternating magnetic fields in the core induce AC voltage onto the secondary winding[1]. The voltage on the secondary winding supplies electrical power to an electrical load.

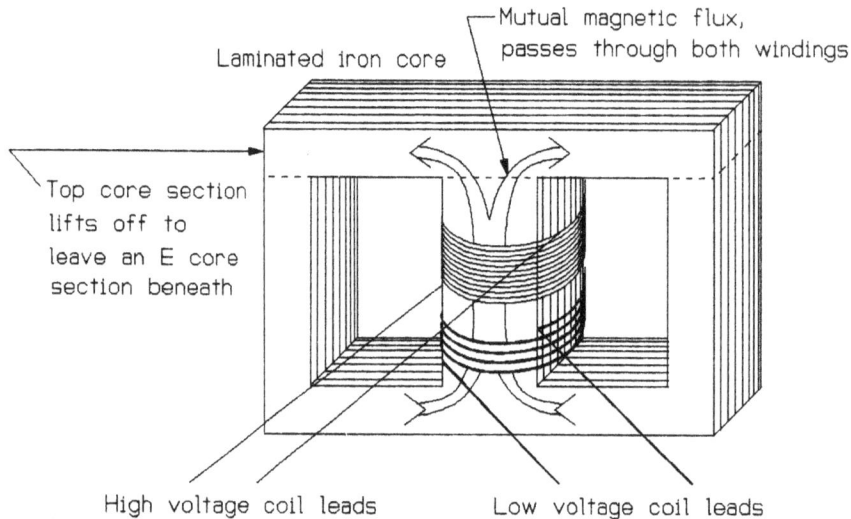

Figure 2-1 E core two winding transformer.

  In a properly made transformer the output voltage is the input voltage times the ratio of the (secondary winding turns)/(primary winding turns). For example, a transformer that received 120 volts AC at the primary winding and produced 12,000 volts AC at the secondary winding would have 100 times more turns of wire on the secondary winding than the primary winding.

Figure 2-2 Transformer schematic symbols.

  Transformers are widely used. They are the usual devices for changing AC voltage levels.

---

[1] ...Magnetic flux is the total flow or quantity of magnetism through a magnetic circuit. A basic phenomenon of electromagnetism is that changing the amount of magnetic flux going through a loop of wire induces voltage on the wire.

One important use of transformers is in electrical power transmission systems. Power station alternators typically produce voltages in the 15,000 volt range. Transformers in power station switchyards step up this voltage to much higher voltages for cross-country transmission. Transmission voltages range from 69,000 to 500,000 volts. Near the customer other transformers step down the high voltages to lower levels the customer can use. Even inside homes there are many transformers in use. Some transformers in the home are used to step 120 volts AC down to low voltage levels. These appear in radios, TVs, and other electronic devices. There are also transformers in the home that step 120 volts AC up to much higher voltages. Some of these step up transformers appear in TVs, microwave cookers, and oil burner ignitors.

*** INDUCTION COIL

The induction coil is a special type of soft iron core transformer. It operates with pulsed DC (rapidly switched on and off DC electricity).

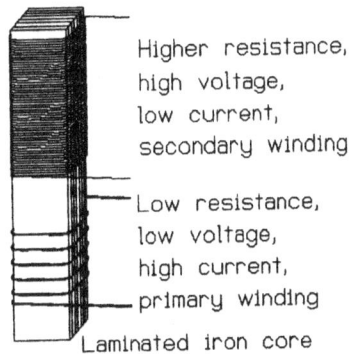

Higher resistance, high voltage, low current, secondary winding

Low resistance, low voltage, high current, primary winding

Laminated iron core

Figure 2-3 Typical induction coil.

The primary winding of an induction coil is a few turns of heavy gauge wire. DC voltage is applied to the primary winding and the electrical current flowing through it rapidly climbs to some high level. As the current increases, the amount of magnetic flux through the core and surrounding air increases. This is shown in Figure 2-4.

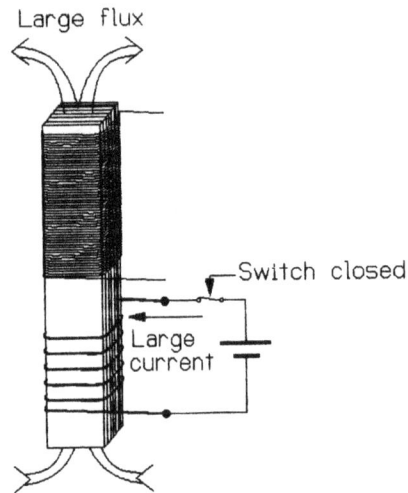

Figure 2-4 Current through primary winding, flux strength increasing.

After the current and flux have risen to a near maximum level, the switch to the input DC voltage is suddenly opened. This is shown in Figure 2-5.

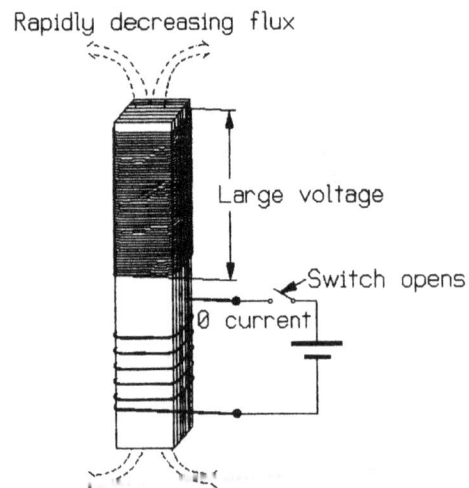

Figure 2-5 Current through the primary winding is interrupted, causing the flux to rapidly decay to zero.

The flux in the core rapidly decays to zero because the core material is "soft iron," a type of steel that does not permanently magnetize. The rapid change of flux induces voltages on the windings. Each turn of wire in each winding has a voltage induced on it by the rapidly changing magnetic flux. The secondary winding has many turns. The voltages induced on each turn add to each other to result in a large voltage surge at the terminals of the secondary winding.

Induction coils are used on gasoline powered engines to provide voltage to the spark plugs. The magnitude of the output voltage surge produced on a car's induction (spark) coil is about 10,000 volts. Higher voltage magnitudes are produced by larger induction coils that have more turns of wire on their output windings.

*** TESLA COIL

Tesla coils are air core transformers specially constructed and tuned to produce high frequency, high voltage AC electricity. The output voltages they produce are always high frequency AC. A typical Tesla coil might produce an output voltage of 200,000 volts AC at a frequency of 400,000 Hz.

Figure 2-6 shows a schematic diagram of a typical electrical arc/capacitive discharge Tesla coil generator. This is similar to the circuit that was built by Nikola Tesla about 100 years ago.

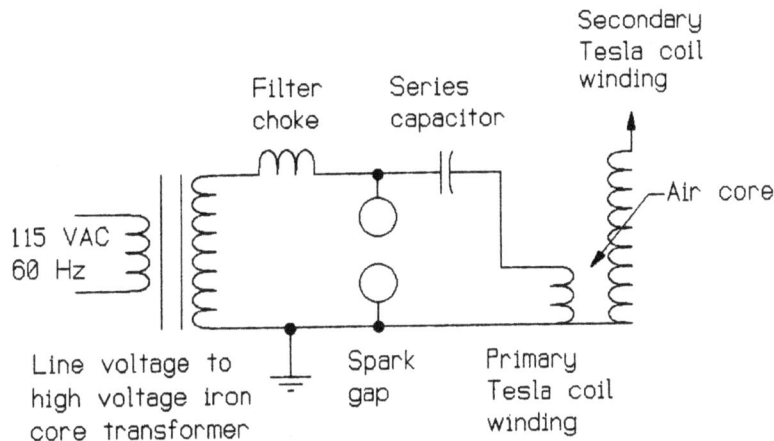

Figure **2-6** Classic Tesla coil circuit.

The basic parts are:

     <u>Line voltage to high voltage iron core transformer</u>. This transformer takes 60 Hz 120 volts AC and steps it up to a high voltage of several thousand volts. The high voltage feeds the spark gap. Even though this chapter is not a construction plan it is worth putting in a note of caution for those considering the building of a Tesla coil. The output voltage of an iron core transformer powerful enough to operate a Tesla coil is definitely lethal. THE PRIMARY CIRCUIT OF A TESLA COIL PRODUCES LETHAL VOLTAGES, even if the high frequency high voltage output of the Tesla coil is safe to touch.

     <u>Filter choke</u>. This reduces the amount of EMI (electromagnetic interference) sent back through the initial transformer. EMI can cause radio and television interference. It is usually just a couple of turns of wire around a ferrite core.

     <u>Spark gap</u>. During Tesla coil operation the arc across the spark gap increases and decreases its resistance in a repeating cycle.

     When the line to high voltage iron core transformer supplies enough voltage to the spark gap it arcs across and acts as a negative resistance (increase in current through the spark gap arc results in a decrease in voltage across the spark gap). The negative resistance of the arc combines with the positive resistance of the series capacitor and Tesla coil primary winding to result in a total circuit resistance that is very near to zero. Circuits that have a zero total circuit resistance are undamped and unstable. The result is high frequency current oscillations through and voltage oscillations across the spark gap, series capacitor, and Tesla coil primary.

     <u>Series capacitor</u>. The series capacitor is a high voltage, low capacitance capacitor that helps set the resonant frequency[2] to the Tesla coil primary. In the past a Leyden jar capacitor was used for this.

     <u>Primary winding</u>. The primary winding of the Tesla coil transformer is a few turns of well spaced heavy gauge wire wound in a coil. A typical primary winding might have a four turns of #14 AWG wire wound with a turn to turn spacing of 1/2". The inductance of the

---

[2]...Resonant frequency is an important concept used throughout electrical work. A circuit made with an inductor and a capacitor in series will have a low impedance to electrical current, if the electrical current is AC of the resonant frequency. If AC of a frequency other than the resonant frequency is applied then the impedance will be greater. The equation for calculating the resonant frequency is:

$$f = \frac{1}{6.28(LC)^{1/2}}$$

f is frequency in Hertz, abbreviated as Hz
L is inductance in Henries, abbreviated as H
C is capacitance in Farads, abbreviated as F

primary winding combines with the capacitance of the series capacitor to act as a bandpass filter for a selected resonant frequency. The resonant frequency effect does two things here. First, it allows only the desired frequency into the Tesla coil transformer. Second, it boosts the voltage on the primary winding. The voltage on the primary winding will be much more than that across the spark gap. The voltage across the Tesla coil primary winding can be high enough so that sparking would occur from turn to turn, if the turns were not spaced far enough apart.

Secondary winding. The secondary winding of a Tesla coil transformer contains more turns of wire than the primary winding. Turns are located so that they are inside the circumference of the primary winding. The gauge of the wire used is smaller than that of the primary winding. There is space between the turns of insulated wire, although not as much as on the primary. A typical winding might use several hundred turns of #18 AWG wire.

The secondary winding of the Tesla coil transformer does two things simultaneously. First, it acts as a normal transformer secondary coil. The primary winding of the Tesla coil transformer produces an oscillating magnetic field that induces voltages on the secondary winding. In normal properly operating transformers, the secondary winding voltage is simply the primary voltage times the number of turns on the secondary divided by the number of turns on the primary. Second, and very important to true Tesla coil operation, the secondary acts as a tuned antenna. It is built as a bandpass filter to allow the same resonant frequency as the primary winding circuit. Nikola Tesla discovered the importance of making the secondary winding with wire of length 1/4 the wavelength of the operating resonant frequency. This is similar to what ham radio operators do when building antennas. To give an example, if the Tesla coil output frequency is to be 400,000 Hz, then the length of the wire in the secondary winding should be:

(1/4)((speed of light)/(frequency)) = 615 feet

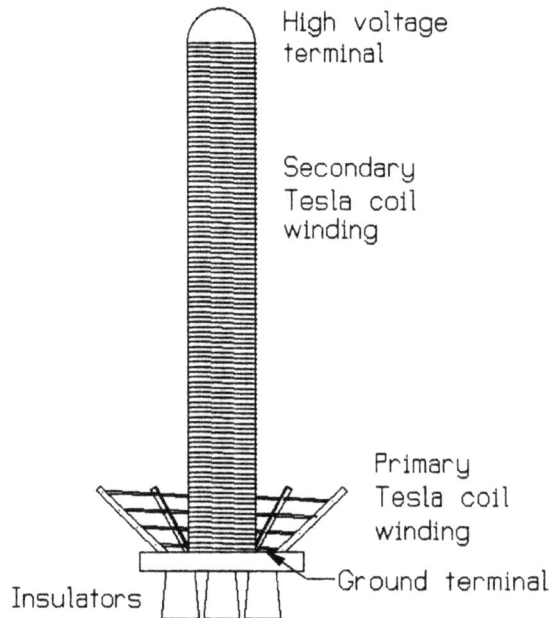

Figure 2-7 Sketch of a typical Tesla coil.

Because a Tesla coil's output voltage is at high frequency, its output current does not penetrate into conductors. If the output of a Tesla coil is connected to a wire, most of the current will be on the outer layers of the wire, the wire's center will not carry much current. If the output of a Tesla coil is connected to a person most of the current goes through the outer layer of the person, along a person's skin. A person's inside would receive little current from the output of a Tesla coil. The technical name for this is "skin effect" (no kidding).

Because of the tendency for currents to go along the outside of a person, it is possible to do spectacular electrical demonstrations with Tesla coils. During demonstrations people have allowed themselves to be exposed to tremendous Tesla coil output voltages without being electrocuted. It should be noted though that the high frequency electric arc from a Tesla coil may cause burns at the point of contact. For example, a person could receive severe burns on a finger tip by receiving Tesla coil arcs to it.

Presently, there aren't many practical uses for Tesla coils. Their main use is in creating spectacular electrical demonstrations.

*** COCKCROFT-WALTON VOLTAGE MULTIPLIER CIRCUIT

A Cockcroft-Walton voltage multiplier circuit produces high voltage DC output from an AC input. It can produce up to one million volts.

The multiplier circuit has two major components, high voltage capacitors and high voltage rectifiers. Multipliers receive AC and then rectify and store it with a series of capacitor/rectifier stages. The more stages there are in the multiplier circuit the more the output voltage is raised.

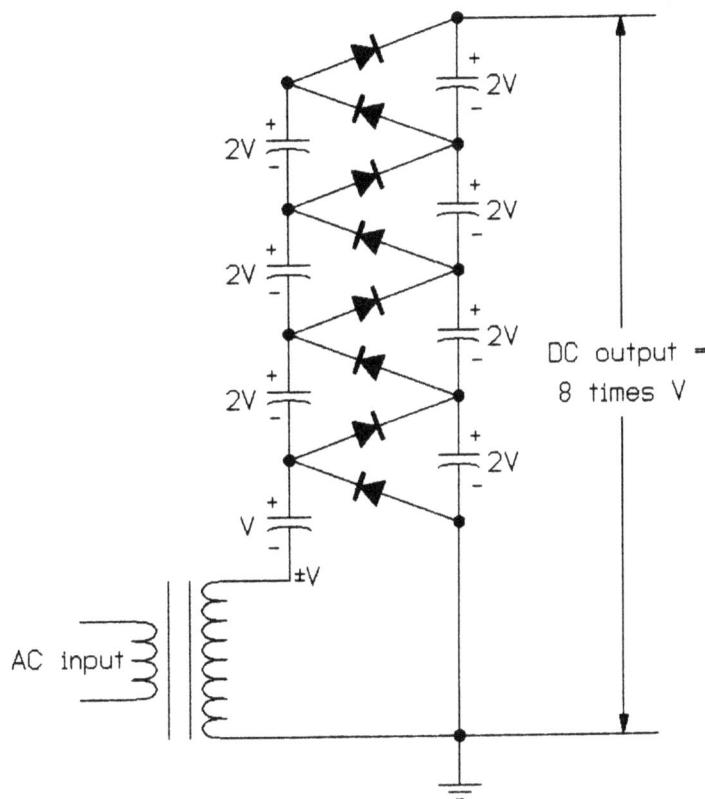

Figure 2-8 Example of a Cockcroft-Walton 4-stage voltage multiplier. Multiplier input AC voltage is V peak AC volts. Output voltage is DC voltage with a magnitude of eight times V.

The capacitors charge to DC voltages as shown in Figure 2-8. The values of the capacitor voltages add to result in a DC output voltage.

One way of explaining the operation of the circuit is to think of the capacitors as accumulators of DC voltage and conductors of AC voltage. Meanwhile, the rectifiers pass current whenever they have a more positive voltage on their anode than their cathode (in other words, when they are forward biased). To reduce capacitive impedance, higher frequency AC supplies are often used.

Cockcroft and Walton originally developed this circuit in the 1930's as a high voltage supply for particle accelerators. They used vacuum tube rectifiers. Now solid state rectifiers are used. One common use of this circuit is in color TVs where a voltage multiplier is used to raise the 10,000 volt output of a flyback transformer to the 30,000 volt needed by the picture tube.

*** MARX GENERATOR

The Marx generator is a circuit that produces a single high voltage surge when it is triggered. Marx generators have been built that can produce surges of millions of volts.

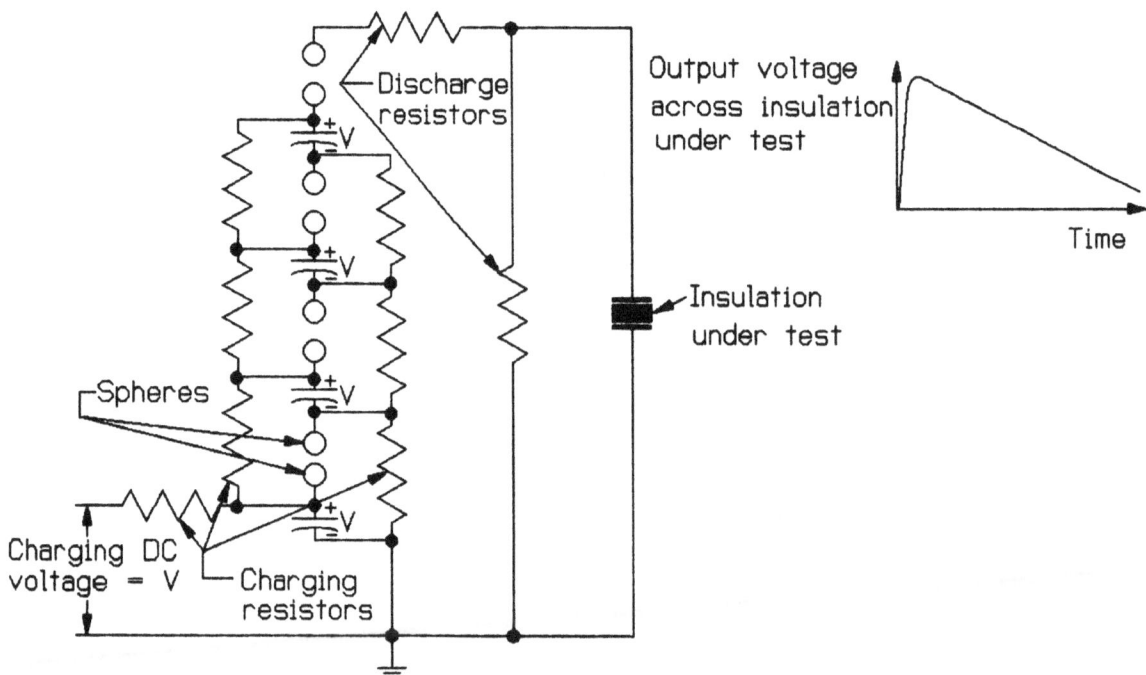

Figure 2-9 Marx generator.

The Marx generator contains stages of capacitors and sphere gaps. During charging the capacitors are charged in parallel to the level of the DC charging voltage. After the capacitors are charged to a desired voltage then the spheres of each pair are simultaneously moved closer to each other. The gaps are set so that the bottom spheres are always slightly closer than the other sphere pairs. As the spheres are moved closer the bottom sphere gap suddenly arcs across. This causes the upper gaps to also arc across. Each arc functions like a closed switch, effectively connecting the capacitors in series. The result is that the Marx generator supplies a single surge of capacitor discharge voltage to its output.

The Marx generator can simulate short duration lightning impulses and longer duration electrical switching impulses seen in electric utility transmission systems. The principle use of the Marx generator is testing the electric breakdown strength of high voltage insulation.

*** VAN DE GRAAFF GENERATOR

Van de Graaff generators produce very high DC voltages but very low currents. They are electrostatic machines.

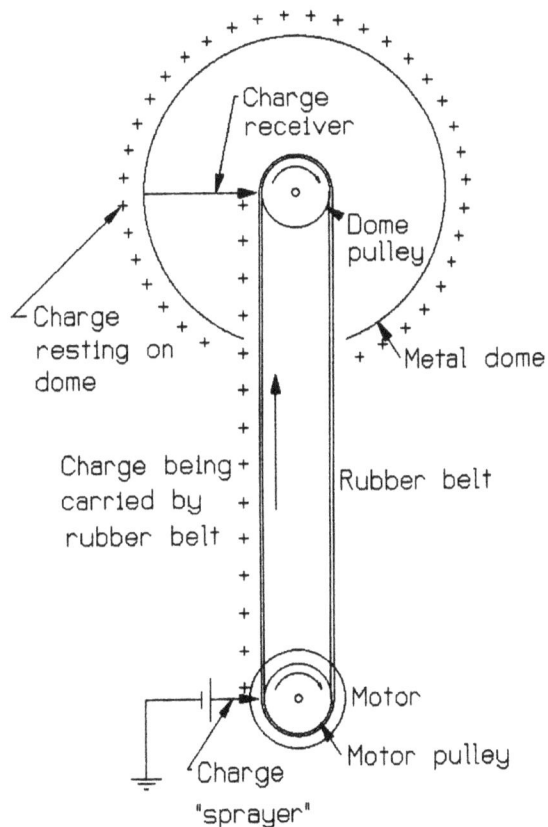

Figure 2-10 Typical Van de Graaff generator.

Van de Graaff generators use a belt to carry electric charge from a DC supply near their motor pulley to a collector in their dome.

At the motor pulley end needle-like points of metal are used to "spray" charge onto the moving belt. Needle-like points are used because they better ionize the air near them. Charge is supplied by a positive or negative DC voltage source between the needle-like points and ground.

The belt is in constant motion. It carries the charges that have landed on it into the dome.

In the dome more needle-like points of metal are used to receive charge from the moving belt. Charges that land on the dome's needle-like points are immediately conducted to the dome. On the dome the charges are as far away as possible from each other. They go there due to the mutual repulsion of like electrical charge. While the dome is becoming charged, the dome's needle-like points are uncharged. Therefore, the dome's needle-like points continue to receive charge from the charged belt. The result is that the dome becomes more and more charged until corona leakage through the air discharges the dome as fast as the belt charges it.

Van de Graaff generators are used in electrical demonstrations and as high voltage supplies for atom smashers. They produce high voltages, but have low current capacities. Because of the low current capacities, smaller Van de Graaff generators can be operated without many safety precautions.

See chapter 11, VAN DE GRAAFF GENERATOR PLANS.

*** BUILDING HIGH VOLTAGE SUPPLIES

If you want to build high voltage sources (other than the Van De Graaff generator), you will need more knowledge than can be gained from this book. I would recommend that you take electrical classes at a vocational school and/or college. Your studies should include hands-on experiments with electrical machinery and transformers. Working with electrical machinery and transformers powered by hundreds of volts will help prepare you for working with higher voltages.

*** REFERENCES

Curtis, Thomas Stanley, *High Frequency Apparatus* (Bradley, IL: Lindsay Publications Inc., 1916). Purchase from Lindsay Publications Inc., P.O. Box 12, Bradley, IL 60915-0012.
    Lindsay has reprinted this book. It is a collection of state of the art 1916 information on the construction of transformers, induction coils, and Tesla coils. Library of Congress number for the original version: QC543 C8.

Pringle, Todd A., *Tesla Coil Handbook* (Jamestown, ND: Todd Pringle, 1993). Purchase from Todd Pringle, 2308 5th Street NE, Jamestown, ND 58401.
    This 60 page booklet contains details and plans for Tesla coils that Pringle built.

3. ELECTRIC SHOCK

What causes electric shock, voltage or current? This question is debated endlessly. The answer is that they both do. Any circuit producing an electric shock has both voltage and current occurring simultaneously. It is impossible to have one without the other.

The amount of electrical current a person receives with an electric shock depends on the magnitude of the voltage supply, the current capacity of the supply, the resistance of the person, and the contact resistance between the voltage supply and the person. Wet, sweaty skin will make a better electric current path than dry skin. The frequency of the voltage may also make a difference, although there is no significant difference between DC voltage and 60 Hz AC voltage.

Tests have shown that people can first feel a tingling from electric currents between .7 and 1.1 milliamps. Although the tingling is not pleasant, it is not particularly dangerous for short periods of time.

Other tests have shown that when people receive currents between 10.5 and 16 milliamps that they lose muscle control. If a person has grabbed an electrically live conductor that supplies him with current of 10.5 to 16 milliamps, he may not be able to let go of it. At these and higher currents the heart may be forced into an erratic and ineffective pumping motion called ventricular fibrillation. When this happens a person is likely to die in 4 to 6 minutes. Tests with monkeys have shown that voltages as low as 45 to 90 volts can cause ventricular fibrillation.

With higher currents, significant resistive heating also takes place. A person receiving a high current shock from something like a high voltage transmission line or the electric chair has his body burned and destroyed by the heat produced by the electricity going through him.

*** REFERENCE

Dalziel, C. F., "Electric Shock Hazard", *IEEE Spectrum*, Vol. 9 (Feb. 1972) pp. 41-50.

4. FUSES AND CIRCUIT BREAKERS

Fuses and circuit breakers are used to protect wiring and equipment from overcurrents.

The most visible use of fuses and circuit breakers is in the home. They are designed to open (interrupt or trip) when the current through them exceeds the ratings of the house's wiring. Without fuses or circuit breakers the currents going through a house's wiring might be allowed to exceed its rating. This would cause the wiring to overheat, eventually destroy its insulation, and perhaps cause a fire. The most important reason for fuses and circuit breakers in your home is to prevent electrical fires from starting in the wiring.

In cars, fuses are used to prevent electrical fires just as they are in homes.

In some electronic equipment fuses are used to protect circuitry from overheating. Fuses are often found in audio amplifiers, oscilloscopes, and multimeters.

*** WHAT ARE FUSES AND CIRCUIT BREAKERS NOT USED FOR?

They are usually not used to protect people from electric shock. The electric current sufficient to cause electric shock is much less that the current typically required to blow a fuse or trip a circuit breaker.

They do not work as direct protection for semiconductor devices. This is because semiconductors fail on overcurrent much faster than fuses or circuit breakers operate.

*** FUSE BASICS

The basic mechanism of every fuse is an electric current carrying element that melts open when too much current goes through it for too long.

When electric current goes through any conductor it heats the conductor. The same is true for the conducting fuse element. It is heated by the passage of current through it, the greater the current the greater the heating. The time of passage of current is also important. The longer the current is applied the more heat energy the fuse element receives. The more heat energy the fuse element receives the greater its temperature will become. A fuse element is heated when it is passing normal currents, but it radiates and conducts away heat fast enough so that its temperature remains low. When a fuse element passes a current that is above its rating it heats up to the point where the element melts and then open circuits by exploding apart. How fast it does this mostly depends on the magnitude of the current. For example, consider a typical single element 15 amp fuse. That fuse can carry 15 amps indefinitely. If the fuse's current were suddenly increased from 15 amps to 150 amps it would blow open in about .3 seconds. If the fuse's current were suddenly increased from 15 amps to 800 amps it would blow open in about .01 seconds.

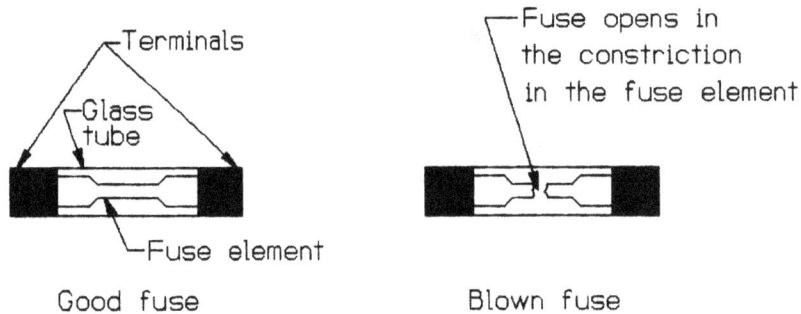

Figure 4-1 Standard single element fuse of the glass cartridge type. Note the constriction in the fuse element, that is where the element resistance is the highest and where the element would melt open on overcurrent.

*** BASIC FUSE TYPES

Fuses come in many different shapes. In the home's fuse box, the screw type and the cartridge type are found. Usually the screw type is used to protect branch circuits that go to parts of the house and the cartridge type is used in the main disconnect. In electronic devices like audio amplifiers small glass cartridge fuses are used. In automobiles small glass cartridge fuses and special automotive plug-in fuses are used.

There are four basic fuse speed types. The standard single element fuse, the time delay fuse, the dual element fuse, and the new semiconductor fuse.

The standard single fuse element has one element that blows open relatively quickly when excessive current goes through it. This fuse is shown in Figure 4-1.

The time delay fuse has a relatively heavy and low melting temperature fuse element. This element must be heated for some time before it melts and open circuits. Often there is also a spring that is part of the fuse element. When the fuse receives overcurrent for long enough the element melts and the spring pulls the connection apart, opening the circuit. These fuses are often used in motor circuits, since motors draw high currents during starting.

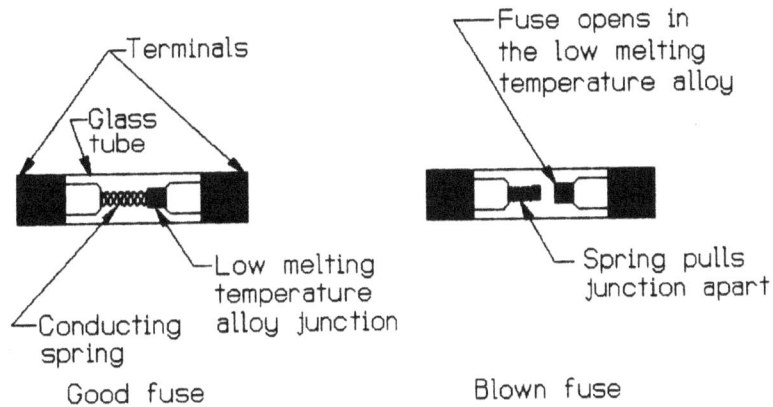

Figure 4-2 Time delay fuse of the glass cartridge type. Note the relatively massive low melting temperature alloy junction, that is where the element resistance is the highest and where the element would melt open on overcurrent.

The dual element fuse has a fuse element for fast action and a fuse element for time delay action. On sudden very high overcurrents the fast action element of the fuse blows. On long enduring overcurrents that are slightly over the fuse rating the time delay element of the fuse blows. These fuses offer the best protection for motors. They don't blow open on the short duration high starting currents, but will blow open if the motor is overloaded for a long time.

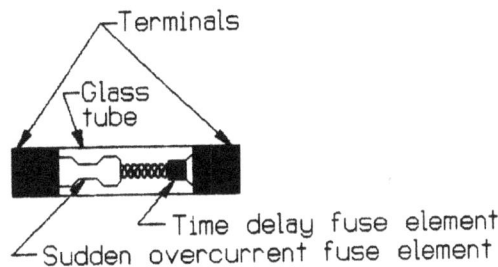

Figure 4-3 Dual element fuse of the glass cartridge type.

There are now semiconductor fuses available. Presently, these are just fast acting fuses similar to single element fuses. Built to blow open as fast as possible, they are still much slower than semiconductor devices. Semiconductor fuses are used in industry to protect large power electronic devices.

*** CIRCUIT BREAKERS

Circuit breakers are spring operated automatic switches that open circuit when the current going through them exceeds a trip current. Circuit breakers can not actuate instantly, it takes time for the mechanisms to move. In speed of operation, circuit breakers are roughly equivalent to time delay fuses.

The most common type of circuit breaker is the thermal trip type. These are used in homes. In the thermal trip circuit breaker there is usually a current carrying bimetal strip. When the bimetal strip carries electric current it heats up. A bimetal strip bends as it is heated. When the bimetal strip bends far enough it mechanically releases the spring operated switch of the circuit breaker, opening the circuit breaker. It is not possible to show the physical construction of a circuit breaker in a simple drawing, there are too many parts. Figure 4-4 shows a rough sketch of the important parts of a thermal trip circuit breaker.

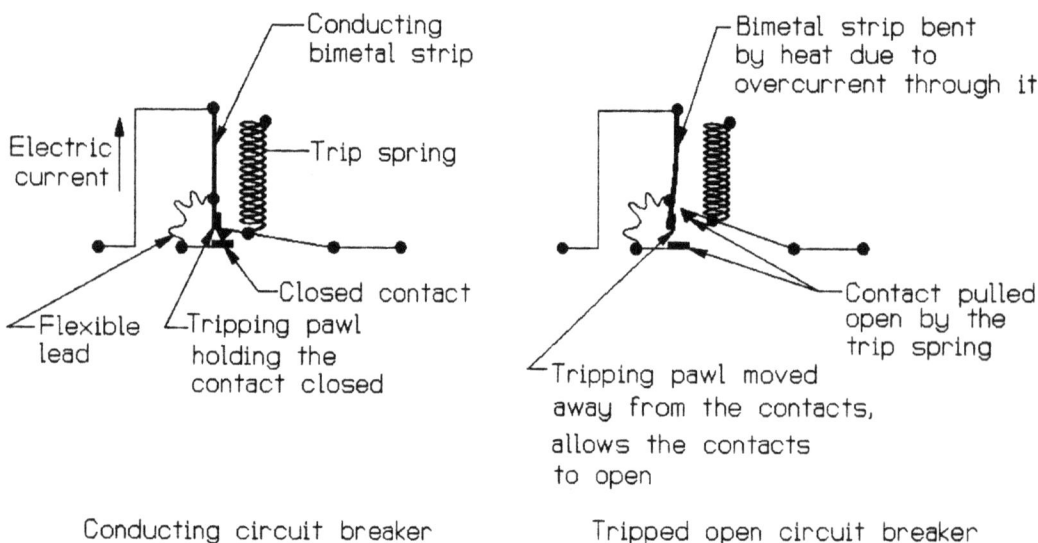

Figure 4-4 Rough sketch of a single-phase thermal trip circuit breaker.

Magnetically operated circuit breakers use a moveable iron core instead of a bimetal strip to trigger the spring operated switch of the circuit breaker. The moveable iron core is forced to move due to its attraction to a magnetic field created by a coil carrying the circuit breaker's current. The moveable iron core is pulled back by a spring as it is magnetically pulled through a motion damping hydraulic fluid. If the electric current is long enough and great enough the moveable iron core travels to a position where it triggers the motion of the spring operated switch of the circuit breaker. New homes often use magnetically operated rather than thermally operated circuit breakers.

There are also remotely operated circuit breakers. Switchyards of power stations and substations use very large circuit breakers that are powered by springs or compressed air. These are made to trip by overcurrents, abnormal voltages, high temperature sensed in a nearby transformer, and many other causes.

*** VOLTAGE AND CURRENT RATINGS

Fuses and circuit breakers have two current ratings. The first and most important is the interrupting current. A 15 amp fuse or circuit breaker should open if the current through it is greater than 15 amps for some period of time. The second current rating is the maximum current that the fuse or circuit breaker can be depended on to withstand and operate properly. For example a typical 15 amp fuse can safely interrupt currents of up to 10,000 amps. If the 15 amp fuse were to somehow suddenly be carrying a current greater than 10,000 amps it might do more than simply blow open. Instead, it might destructively explode apart and reconnect the circuit by arcing through the air or across the fuse holder that the fuse was mounted on.

Fuses and circuit breakers also have voltage ratings. Those used in house wiring are usually rated to 600 volts. If the voltages used are higher than rated, problems may arise. First, a high voltage might simply arc from the fuse or circuit breaker to ground. Second, if the fuse or circuit breaker has to interrupt a voltage greater than its rating, the voltage may cause a continuous arc from terminal to terminal in the fuse or from contact to contact in the circuit breaker. This arcing may continue for some time or may destructively explode apart the fuse or circuit breaker.

*** ADVANTAGES AND DISADVANTAGES OF FUSES AND CIRCUIT BREAKERS

Circuit breakers have the following advantages over fuses:
1) It is easier to reset a circuit breaker than replace a fuse.
2) In homes the use of circuit breakers eliminates the danger of a stupid person using pennies or oversized fuses to replace blown fuses.
3) Circuit breakers can often be used as switches.

Fuses have the following advantages over circuit breakers:
1) Fuses are cheaper than circuit breakers, provided that they are not blown often.
2) Fuses are faster operating than circuit breakers.
3) Fuses are not affected by vibration as much as circuit breakers.
4) Fuses are smaller and lighter than circuit breakers.
5) In the author's opinion, fuses are more reliable than circuit breakers.
6) Fuses are convenient for the experimenter who is testing out different variations of a circuit.

*** SELECTING A MOTOR FUSE OR CIRCUIT BREAKER

Be certain that the voltage and maximum current capabilities of your supply do not exceed those of the selected fuse or circuit breaker.

In house wiring #12 AWG copper wire is rated to continuously carry up to 20 amps. The fuse or circuit breaker protecting it should have a current interrupt value of 20 amps or less. #14 AWG copper wire used in house wiring is rated to continuously carry 15 amps. The fuse or circuit breaker used to protect it should have a current interrupt value of 15 amps or less. For other wiring sizes, select a fuse with a current interrupt value less than or equal to the wire's current rating. Wire current ratings can be found in the National Electrical Code (see page 131) and many other electrical handbooks.

Motor fuse or circuit breaker selection information should be obtained from the motor manufacturer and/or fuse or circuit breaker manufacturer. However, there are some rules of thumb.

1) With single element fuses: Select a fuse that has a current interrupt value equal to or less than three times the rated full-load current of the motor.

2) With dual element fuses: Select a fuse that has the largest delayed current interrupt value less than or equal to 1.25 times the rated full-load current of the motor.

3) With circuit breakers: Select a circuit breaker that has the largest current interrupt value less than or equal to 1.25 times the rated full-load current of the motor.

If experimental equipment is being built, for the best protection use fuses with the smallest current interrupt value that will allow proper equipment operation.

Coordination of fuses and circuit breakers is required in the building of electrical systems. In a properly coordinated system the fuse or circuit breaker closest to a short circuit blows or trips open before a farther away fuse or circuit breaker. This is seen in house wiring. If I have a sudden 100 amp short circuit in the kitchen branch circuit of my house the 15 amp kitchen fuse blows, but the 60 amp fuse in the main disconnect is still good. That way there is still electricity going to the other branch circuits in the house. If I had previously incorrectly replaced my 15 amp kitchen fuse with a 90 amp fuse then my 100 amp kitchen short circuit would have blown the main fuse and shut down the whole house. That would have been bad coordination.

*** GROUND FAULT INTERRUPTER

Ground fault interrupters are related to circuit breakers, but they have a different purpose. They protect people.

In a house a ground fault occurs when part or all of the electrical current to a piece of equipment returns to the service entrance ground by a path other than the wiring. An example of this is shown in Figure 4-5.

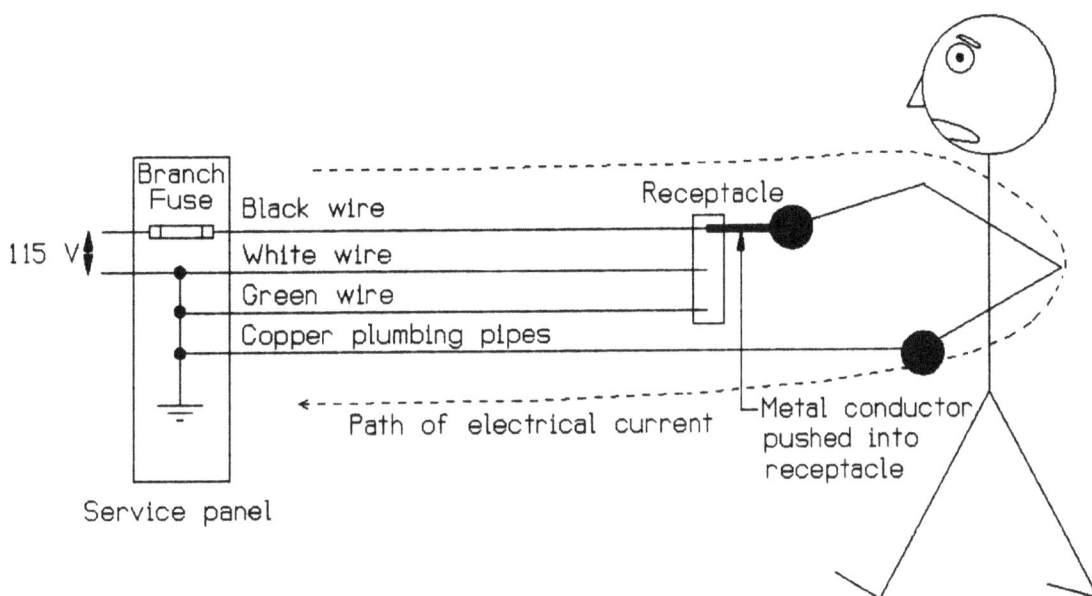

Figure 4-5 Without a ground fault interrupter, this person is receiving a serious, possibly fatal, electric shock. In one hand he holds a metal conductor inserted into the hot side of an electrical outlet while his other hand holds onto a faucet connected to the grounded copper pipe plumbing system. The shock received here is the maximum possible, the full line voltage of 120 volts. If this person received a shock from an appliance with failing insulation he might receive a lower voltage shock. Notice that electrical current is going out via the black wire, but is not returning by the white wire.

A ground fault interrupter determines that a ground fault is occurring by sensing that the currents going through the current carrying wires are not equal. The two wires (or three if three-phase is being used) go through a differential current detector. This is a ring of magnetic steel transformer laminations. The wires act as a one turn transformer primary winding. A secondary winding is also wound around the ring. When there is a ground fault some current returns to the service panel ground point without going through the white wire. This means that unequal currents are flowing in the white and black wires. When unequal currents are flowing, the secondary coil has voltage induced on it. This voltage then triggers a spring powered switch that disconnects the circuit.

A typical ground fault interrupter will trip when there is a difference in the wire currents of 5 milliamps. The time required for the ground fault interrupter to operate is about 25 milliseconds.

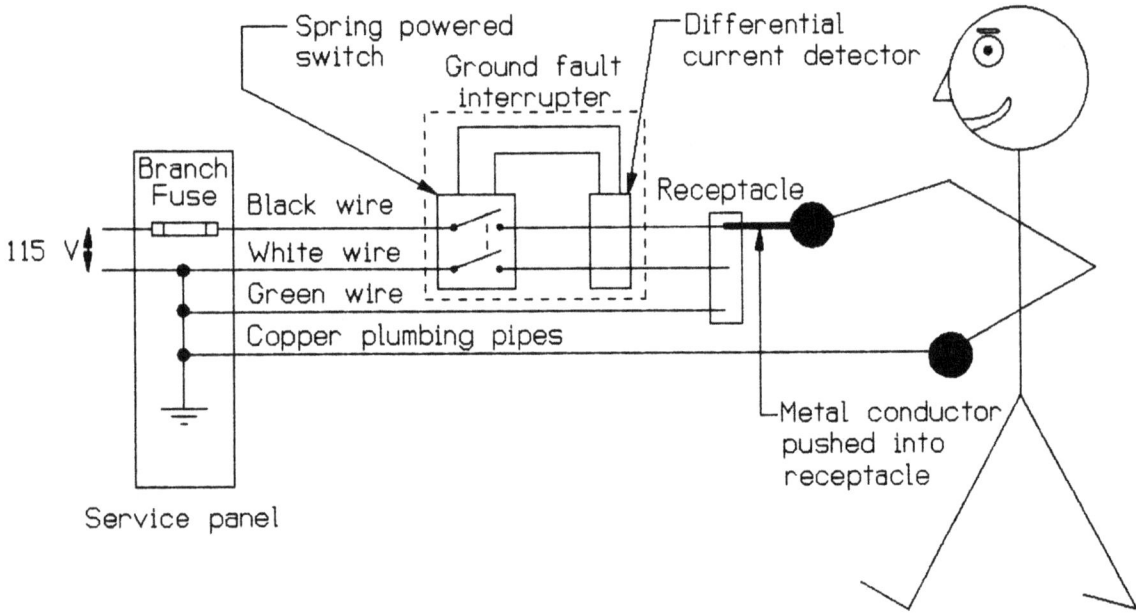

Figure 4-6 With the ground fault interrupter this person is being protected from receiving a continuous ground fault electric shock. The initial current here was the same as in Figure 4-5. However, this current was interrupted in about 25 milliseconds, before serious damage was done to the person.

*** MANUFACTURER'S CATALOGS

The following manufacturers offer free informative and detailed catalogs.

For fuses:
Cooper Industries
Bussmann Division
St. Louis, MO 63178-4460
http://www.bussmann.com/

For circuit breakers:
Eaton Corporation
1-800-356-1243
http://web.eaton.com

For ground fault interrupters:
Leviton Manufacturing Co., Inc.
59-25 Little Neck Pkwy.
Little Neck, NY 11362
718-229-4040
fax 800-832-9538
http://www.leviton.com

5. SINGLE-PHASE TO THREE-PHASE IDLER MOTOR CONVERTER

*** WHAT ARE SINGLE-PHASE AND THREE-PHASE ELECTRICITY?

Single-phase is the type of AC (alternating current) electricity used in homes and in other small power applications. It works well with lighting and heating. It will also power single-phase motors. Single-phase electricity uses two or three current carrying wires and a ground wire. A black wire and a white wire are used to carry current in 120 volt wiring. Red, black, and white wires are used to carry current in 240 volt wiring.

Three-phase is the type of AC electricity used by the large scale electrical user. It is preferred by the electrical generating utilities because it can be more efficiently generated and transmitted than single-phase. It is preferred by the industrial user because it allows the use of three-phase induction motors. The three-phase induction motor is more rugged and less expensive than a similar capacity single-phase induction motor. Three-phase electricity uses three current carrying wires.

*** WHERE TO USE A SINGLE-PHASE TO THREE-PHASE IDLER MOTOR CONVERTER?

The principle use for a single-phase to three-phase idler motor converter is to power small three-phase motors. The three-phase motors might be on: blowers, small compressors, drill presses, grinders, hoists, lathes, milling machines, pumps, punch presses, and rams.

A single-phase to three-phase idler motor converter may also be used to power some three-phase electronic equipment.

*** WHAT IS A SINGLE-PHASE TO THREE-PHASE IDLER MOTOR CONVERTER?

It is a three-phase motor powered on two of its input terminals by single-phase AC. It runs with nothing connected to its shaft. As it operates it generates two more phases of electricity. These phase voltages appear on the motor's third terminal. The generated three-phase may be used to power three-phase motors and other equipment.

It is the least expensive way to convert single-phase to three-phase, costing much less than commercially available converters.

The AC it produces is electrically clean. The three phase-idler motor converter will not produce radio and television interference.

*** WHO SHOULD MAKE A SINGLE-PHASE TO THREE-PHASE IDLER MOTOR CONVERTER?

The maker of the three-phase idler motor converter should know basic wiring techniques. If the maker is uncertain of his electrical abilities it is suggested that an electrician inspect his work, before power is applied.

The maker should have the mechanical ability to create a mounting base for the three-phase idler motor converter and three-phase outlet. Also he should be able to make a shaft guard to protect people and equipment from coming in contact with the three-phase idler motor converter's shaft.

*** WHAT IF YOU NEED A CONVERTER, BUT HAVE DECIDED AGAINST MAKING YOUR OWN?

There are companies that manufacture single-phase to three-phase converters. Your local electric motor repair shop may have a dealership for some types.

A few manufacturers are:

ARCO ELECTRICAL PRODUCTS CORP.
2325 E. Michigan Rd.
Shelbyville, IN 46176-3400
1-317-398-9713
http://www.arco-electric.com

GWM CORP.
1-800-437-4273
http://www.gwm4-3phase.com

KAY INDUSTRIES, Inc.
604 N. Hill St.
South Bend, IN 46617
1-800-348-5257
http://www.kayind.com

STEELMAN INDUSTRIES, Inc.
2800 Hwy 135
Kilgore, Texas 75662
1-903-984-3061
http://www.capacitorconvertors.com/rotary.html

*** WHAT IS THE BASIC PRINCIPLE OF THE THREE-PHASE IDLER MOTOR CONVERTER?

The heart of the three-phase idler motor converter is a three-phase motor with nothing connected to its output shaft. It is called the idler motor. During operation the idler motor is powered by single-phase input power. The rotating rotor of the idler motor induces voltages in its windings so that two more phases of electricity are produced. The result is that there is three-phase AC as measured across the converter motor's three terminals. The three-phase AC across the idler motor's three terminals may now be used to start and run other three-phase motors and three-phase loads.

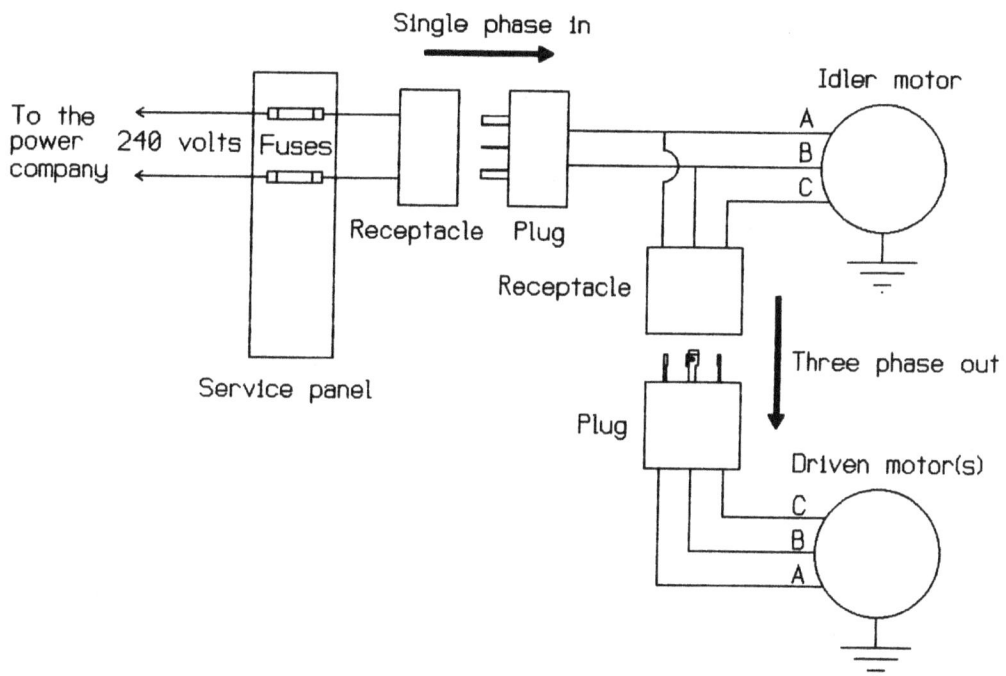

Figure 5-1 Circuit diagram of the single-phase to three-phase idler motor converter.

The idler motor is started spinning by pulling on a rope wrapped around the shaft. The procedure is like the one used to start old outboard motors. The direction of rotation of the idler motor is important. The direction of rotation of the idler motor converter determines the direction of rotation of the driven motor(s). Once the idler motor is spinning and the user is safely away from the idler motor's shaft the user quickly plugs the single-phase input plug into a single-phase outlet. The idler motor then rapidly accelerates to its running speed. Once the idler motor is running steadily then the driven motors may be connected. If more than one motor is to be driven they should be started one at a time, not all at once. Each driven motor should start and rapidly accelerate to speed. The driven three-phase motors may now be used.

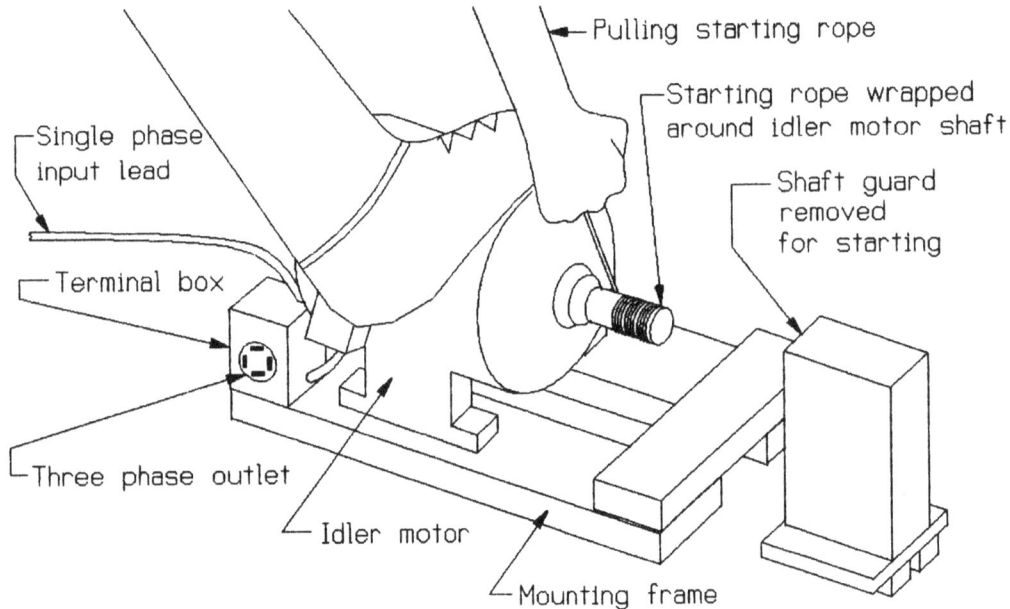

Figure 5-2 A single-phase to three-phase idler motor being started.

Once each driven three-phase motor is spinning at its rated speed it will also function as a three-phase idler motor converter, if it is not connected to a mechanical load. It may be possible for the driven motor to operate without the idler motor converter after it has been started. However, it is usually better to leave the three-phase idler motor converter connected in the circuit. Leaving it connected will give the driven motor an output power capacity closer to its rating.

*** PROCEDURE FOR MAKING A SINGLE-PHASE TO THREE-PHASE IDLER MOTOR CONVERTER

1) Verify that the idler motor converter will work in your application.

    a) The three-phase load you wish to drive should require voltage in the 208 to 240 volt range. Check the load's nameplate to see what its voltage requirement is. The load may be capable of more than one input voltage. For example, 230/460 volt motors are common. If you have one of these dual voltage motors be certain that its terminal leads are connected so that it can receive input voltage in the 208 to 240 volt range.

b) The largest driven motor should not be over 3 HP. The total horsepower of all the driven motors should not exceed 6 HP. Larger idler motor converters will work with much larger loads. However, larger idler motors are hard to start with a rope.

c) In the machine driven by a three-phase motor, be certain that a single-phase motor can not be easily substituted. The considerations are mechanical dimensions and cost.

2) Select a 1725 rpm three-phase induction motor from the following chart.

Largest Driven Motor (HP)	Sum of Driven Motors (HP)	Minimum Idler Motor (HP)
1/3 or less	2/3 or less	½
½	1	¾
1	2	1 ½
2	4	3
3	6	5

3) Buy a second hand three-phase induction motor at an electric motor repair shop or a junk yard. The motor should have a nameplate rating of 208 to 230 volts, at about 1725 rpm. It is possible to use higher speed 3450 rpm motors, but they may be difficult to start. If the horsepower size selected from the above chart is not available then choose the next larger size.

If possible follow the "START-UP PROCEDURE" and start the idler motor on 240 volts single-phase where you are buying it. Some electric motor repair shops will be willing to let you do this.

If you do not have the opportunity to test that you can rope start the motor you are buying, another rule of thumb is that a person usually can not rope start a motor that he can not pick up and carry.

4) Buy wire, socket, plugs, and fuses or circuit breakers at a hardware store. If the user desires, switches may be inserted on the input and output of the converter.

5) It is important that the idler motor and its loads be protected by fuses or circuit breakers. If the electric utility interrupted power for more than a short time the idler motor would coast to a stop. When the electric utility returned power it would be feeding electricity into a stationary idler motor. The idler motor would not start by itself. It would continuously draw a starting

current that might be 6 times larger than that of the idler motor and load motor's nameplate rated current. If those currents were not soon reduced, the idler motor and any load motors it is driving would be destroyed.

An input fuse or circuit breaker should be used to protect the idler motor and its loads. The fuse or circuit breaker should not blow or trip when the high starting currents last for about 2 to 3 seconds during the starting of the idler motor and later of each of its load motors. However, the fuse or circuit breaker should blow or trip if the starting current lasts for more than about 10 seconds.

If your electrical service uses fuses, a time delay or dual element fuse should be used. Circuit breakers that are commonly used in modern electrical service entrances have a built in time delay in their operation. The smallest fuse or circuit breaker that allows the starting of the idler motor and its loads should be used.

To select the size of your input fuse or circuit breaker use the following procedure.
1) Sum the rated nameplate currents of the idler motor and the load motors.
2) Multiply the sum by 1.25. Select the largest time delay or dual element fuse or circuit breaker that is less than or equal to the product of this multiplication.

6) Assemble the circuit. Do not apply power to the converter until it is assembled and ready for starting. If you are not certain that your wiring is correct have it checked by an electrician, have the electrician connect the single-phase outlet to the distribution box and witness the first starting of the idler motor converter.

*** TESTING YOUR SINGLE-PHASE TO THREE-PHASE IDLER MOTOR CONVERTER

FIRST TEST: NO-LOAD TESTING

a) Start your converter using the starting procedure.

b) Measure the line to line output voltages with an AC voltmeter. They should be equal or almost equal to the single-phase input voltage.

c) Allow the converter to run unloaded for several hours. During this time the idler motor will increase its temperature. It should not get too hot to touch or produce a burning insulation smell.

SECOND TEST: LOAD TESTING

a) Start the converter using the starting procedure.

b) Connect loads to the converter output one at a time. Allow each driven induction motor time to accelerate to operating speed before switching in the next.

c) Allow the converter to run loaded for several hours. During this time the idler motor and the load motors will increase their temperatures. Pay attention to the temperatures of the idler motor and the load motors. Most modern induction motors can withstand an increase in temperature of up to 40°C above room temperature. This means a motor's case can go up to a maximum of about 140°F. If you can rest your hand on the idler motor and the load motors, your motors are not being overheated. If you can not rest your hand on them there may or may not be a heating problem. If there is a burning insulation smell at any time there is an over heating problem.

d) If there is no overheating problem, that's great. You now have a functioning converter.

e) If there is a heating problem, it will be necessary to reduce the load on your converter.

*** CASCADING IDLER MOTORS

It is possible to cascade idler motors. A small idler motor converter's three-phase output can be used to start a larger idler motor. Then the capacity of the converter system is determined by the sum of the two idler motor horsepowers. However, the starting of two or more idler motors requires a more complicated starting procedure.

Figure 5-3 Photograph of a single-phase to three-phase idler motor converter. Nameplate data on the motor: 3 HP, 220/440 VOLTS, 8/4 AMPS, 3 PHASE, 1725 RPM, 40°C TEMP RISE, Frame 225.

START-UP PROCEDURE, IDLER MOTOR CONVERTER
post a copy next to the idler motor converter

1) The input power lead to the idler motor must be unplugged.

2) Unplug the output power loads from the idler motor, or turn off each load at its on-off switch.

3) Remove the shaft guard. The shaft should not be spinning.

4) Wrap the starting rope around the shaft of the idler motor converter. The rope must slip off the shaft at the end of pull starting. DO NOT TIE IT TO THE SHAFT. Wrap the rope clockwise or counterclockwise as seen from the shaft end of the motor and then consistently wrap the rope around the shaft that way for each starting. The way the rope is wrapped around the shaft determines the idler motor rotor rotation direction. The idler motor rotor rotation direction determines the generated electrical phase sequence which then determines the driven motor rotation direction.

5) Pull on the starting rope to start the rotor of the idler motor spinning.

6) Quickly plug the input power lead into its outlet. The person who pulls the starting rope should be the same one who plugs in the single-phase plug. This reduces the possibility that a person will be accidentally wound into the idler motor by the starting rope when the motor starts.

7) The rotor of the idler motor should accelerate to a steady speed in a few seconds. If it doesn't then shut off the input power and wait for its rotor to stop spinning. Then repeat steps 4) to 6). Be certain to pull harder on the rope in step 5).

8) Once the idler motor is up to speed, carefully put the shaft guard in place.

9) Turn on the switches to the three-phase loads or plug in the three-phase output lead.

TURN-OFF PROCEDURE, IDLER MOTOR CONVERTER

1) Unplug the single-phase input power.

2) Disconnect the three-phase loads.

6. ASYNCHRONOUS ALTERNATOR

This describes a method of AC electric power generation that uses an asynchronous induction alternator. Information is provided to size one of these alternators to fit the capabilities of an available prime mover[1]. Circuit diagrams and start-up procedures are provided. Locations and sizes of fuses and circuit breakers are suggested.

The asynchronous induction alternator method described assumes that the user's electric system is connected to an electric utility. It also assumes that the user has a prime mover to turn the asynchronous alternator.

*** WHO IS THIS FOR?

This is meant for a person who is interested in producing some of his electric power while still connected to an electric utility system.

It is primarily for the home owner or small scale power user. However, it will be of interest to larger scale power users as well.

*** WHAT IS THE ELECTRIC UTILITY GRID?

In electrical work, "connected in parallel" means connected like terminal to like terminal. Car batteries are sometimes connected in parallel to start cold engines. Paralleling the batteries increases the current and power available to a car's starter motor.

Parallel power generation is the paralleling of the output of an alternator to other alternators. This is done by electric utilities. Most electric utility alternators in the U.S. are connected in parallel. The utilities use meters to measure how much power each utility supplies. The parallel combination of electric utility alternators is called the "electric utility grid."

---

[1]...A prime mover is the engine or machine that turns an alternator or generator. Some examples of prime movers are water wheels, steam turbines, and internal combustion engines.

*** SYNCHRONOUS ALTERNATOR FOR PARALLEL POWER GENERATION

Almost all electric power utilities use synchronous alternators to generate electric power. The alternators are driven with steam turbines (steam from coal, oil, or nuclear energy), water turbines (water from dams) or gas turbines (jet engines).

Utilities carefully set their synchronous alternator rotor speeds, prime mover power outputs, and alternator field voltages (also called excitation voltages) so that their alternator outputs have the desired 60 Hertz frequency, voltage output, current output, and power output.

Standby generators[2] are readily available to the consumer at hardware stores, electrical equipment dealers, and some department stores. Almost all standby generators are synchronous alternators. It is possible for the home owner to parallel the output of a standby generator to his utility company's electric system. However, it is not easy! If the home owner's synchronous alternator has an incorrect speed or output voltage there could be problems when the home owner's synchronous alternator's output is connected. In a worst case scenario the wiring, alternator or its prime mover could be damaged or destroyed.

*** ASYNCHRONOUS INDUCTION ALTERNATOR

The asynchronous induction alternator is simply an induction motor that is driven faster than its synchronous speed by a prime mover.

An asynchronous induction alternator is different from a synchronous alternator in several significant ways:

1) It does not have to operate at a fixed speed. Its operating speed can vary about ±1%.

2) It is not necessary to apply DC field voltage to an asynchronous induction alternator. Therefore, no field control system is needed for an asynchronous alternator.

3) An asynchronous alternator will not operate well by itself without being paralleled to an operating electric grid system. (It is possible to operate an asynchronous alternator isolated from a power system by paralleling it to a properly sized capacitor bank. However, capacitor banks are expensive and asynchronous alternator systems using them are unstable.)

---

[2]...The word generator technically means a creator of DC electricity. The word alternator means a creator of AC electricity. However, the word "generator" is commonly used for standby alternators, so it is used here.

Induction motors have several notable speeds. Rated full-load speed is the speed on the motor's nameplate. It is the speed that the motor will turn when it is producing full torque at rated input voltage. No-load speed is the speed that the motor runs when it has no mechanical load connected to its shaft and rated input voltage is applied. Synchronous speed is a few tenths of a percent faster than the no-load speed. It is the speed at which a theoretical frictionless motor would run if no mechanical load was connected to it.

An induction motor that is used as an asynchronous alternator has a rated full-load speed. When the alternator turns at that speed and is connected to its rated input voltage, it will produce its maximum electric output power. Speeds beyond that will overheat and damage the induction asynchronous alternator.

The following chart gives typical speeds for induction motors used as asynchronous induction alternators. The input frequency is the standard U.S. 60 Hz.

NORMAL INDUCTION MOTOR OPERATION		SYNCHRONOUS SPEED (rpm)	ASYNCHRONOUS INDUCTION ALTERNATOR OPERATION RATED FULL-LOAD SPEED (rpm)
NAMEPLATE FULL-LOAD SPEED (rpm)	NO-LOAD SPEED (rpm)		
850	898	900	956
1140	1197	1200	1267
1725	1795	1800	1882
3450	3590	3600	3764

*** CONNECTING AN ASYNCHRONOUS ALTERNATOR TO THE ELECTRIC UTILITY GRID

Connecting an asynchronous alternator to a home's electric system gives the home owner parallel power generation. The amount of power an asynchronous alternator can supply depends on the power input to the alternator and the alternator's capacity. An alternator may supply only part of the power to the user's home or may totally supply the home and send surplus power to the electric utility.

A sketch of a small simple asynchronous alternator parallel power system for the home follows.

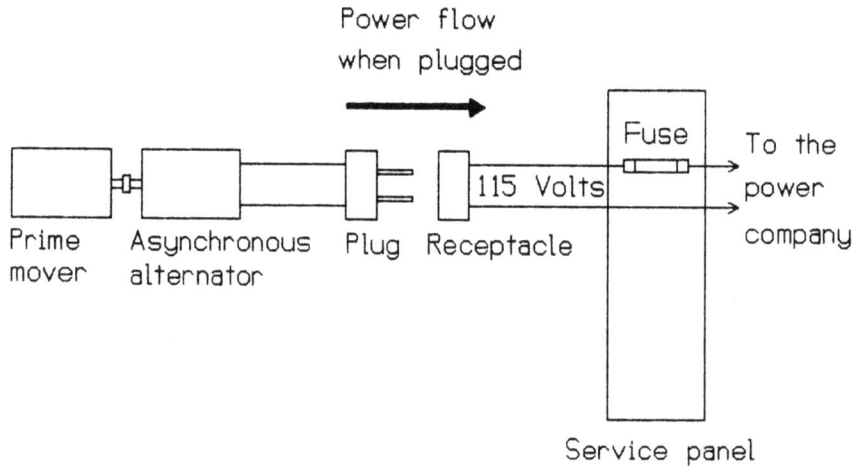

Figure 6-1 Asynchronous alternator parallel power generation system.

*** HOW MUCH POWER CAN YOU MAKE?, HOW MUCH MONEY WILL YOU SAVE?

The amount of electric power made depends on the power your prime mover produces and the efficiency of the asynchronous alternator.

As a rough estimate, the electrical power production will be about 60% of the mechanical output power of the prime mover.

For example, if your prime mover produces 1 horsepower of mechanical power you could expect to produce about (60/100) x 1 x 746 = 448 watts of electrical power. If the utility company cost rate of electricity is $.11 per Kilowatthour then the money saved by generating power from your 1 horsepower prime mover would be (448/1000) x .11 = $ .049 per hour.

*** WHAT TYPES OF PRIME MOVERS WORK WELL?

The prime mover must be able to rotate the asynchronous alternator continuously at a rpm that is slightly greater than synchronous speed.

Water wheels with suitable rpm increasing gearing and water turbines are good prime movers for asynchronous alternators. Wind mills because of their unsteadiness in operating speed and power output are not good prime movers.

The reader is referred to chapter 1, "SAVE MONEY GENERATING ELECTRICITY?", to determine if he has a water power source sufficient to generate a practical amount of power. That chapter also evaluates a user's sources of wind power and solar power.

Another potential power source is a natural gas powered engine. Those with a natural gas well on their property might consider using a natural gas powered engine as their prime mover.

It is suggested that the average mechanical output power capacity of the prime mover be at least a continuous 1/8 horsepower.

*** WHAT ABOUT LEGALITIES?

The watthour meter that measures a home's electric power usage can be slowed or even run backward by power generated by parallel power generation.

If parallel power generation is used to the extent that the meter usually runs backward, the monthly reading of the watthour meter will be negative. One should not expect that he can simply send electricity bills to his electric utility company at the same billing rate the utility uses. Utilities won't pay that rate. They would at least expect some compensation for maintaining their "power available on demand" electric system and then would pay at some lower rate.

There are federal, state, and local laws regulating power flow and practices. The federal Public Utility Regulatory Policy Act (PURPA) of 1978 encourages small scale power generation. Some states have similar encouragements. Investigate and consider the legalities of producing power before doing parallel power generation.

*** ASYNCHRONOUS ALTERNATOR TEST DATA AND RESULTS

Unfortunately, the author does not have a prime mover powerful enough to power a reasonably sized asynchronous alternator. However, the electrical knowledge required for sizing and installing an asynchronous alternator is almost identical to that for sizing and installing an induction motor. The author has done many of these.

This section provides actual electrical data on a 1/4 Hp single-phase induction motor that has been paralleled to the electric utility grid from an electrical laboratory.

For testing the prime mover was a DC electric motor. The DC motor received electric power from the electric utility grid. Of course, using electric power to generate electric power does not result in a net power generation. It results in a net power usage. The DC motor was used only as a convenient laboratory prime mover.

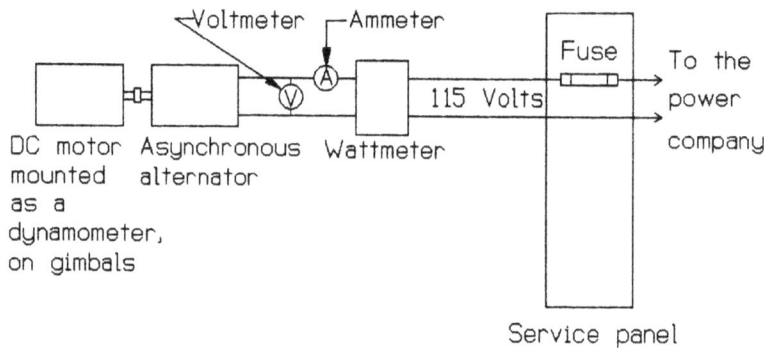

Figure 6-2 Sketch of the laboratory asynchronous alternator test setup.

The nameplate data for the 1/4 horsepower motor is:

GE AC Capacitor Start Motor
MOD 5KC35KG166

1/4 HP	Frame 56C	RPM 1725
PH 1	S.F. 1.35	Temp Rise 40°C
V 115	A 5.2	CODE-M
CY 60	TIME RATING CONT	

Serial Rating TSD

Tests were run with different mechanical power inputs to the 1/4 HP motor. The most significant tests were: 1) When the power output of the asynchronous alternator was zero (no power in or out of the asynchronous alternator). 2) When the asynchronous alternator produced its maximum rated output power.

With the electric power in and out of the 1/4 Hp asynchronous alternator 0 watts the following was observed: AC line voltage = 115 volts AC at 60 Hz, speed = 1800 rpm, mechanical torque turning the asynchronous alternator = .7 Ft.Lbs., current into the asynchronous alternator = 4.2 amps. Note that there is current and voltage applied to the asynchronous alternator, though there is no power going in or out of it. This is possible because of the phase shift of the current relative to voltage. After running for an hour, the asynchronous alternator warmed to an acceptable operating temperature.

The mechanical input power was increased to the 1/4 Hp asynchronous alternator until the current from it was its rated nameplate maximum current. The following was observed: voltage = 115 volts AC at 60 Hz, speed = 1850 rpm, mechanical torque turning the asynchronous alternator = 1.2 Ft.Lbs., current into the asynchronous alternator = 5.2 amps, electric power out to the grid = 145 watts. After running for a long time the motor warmed to a greater temperature than it had in the previous test. While running at full load it was possible to lay a hand on the stator, but it was not comfortable to do so. A thermocouple temperature probe measured the stator temperature as 41°C, an acceptable temperature rise. Calculations show that at full load the efficiency of this asynchronous alternator is 51%.

The results of these tests could be scaled for larger or smaller motors being used as asynchronous alternators.

*** THREE-PHASE MOTOR AS AN ASYNCHRONOUS ALTERNATOR

A three-phase motor can be efficiently used as an asynchronous alternator for sending three-phase electric power to the electric utility grid. However, most small scale users do not have three-phase.

It is possible to use a three-phase motor as an asynchronous alternator to create single-phase electricity. However, the efficiency is much lower this way than with a single-phase motor. Single-phase electricity should be made with asynchronous alternators made from single-phase motors.

*** CONSTRUCTION HINTS

Provide ample ventilation for the asynchronous alternator. A cooler asynchronous alternator will last longer.

Build a covering over or fence around the asynchronous alternator and prime mover that will prevent people and animals from accidentally contacting moving parts or electric circuitry.

Use fuses and/or circuit breakers to protect the asynchronous alternator.

Take care that the asynchronous alternator shaft does not receive too much radial force from its connection to the prime mover. Misalignment of a prime mover to asynchronous alternator coupling or over tightness of a belt and pulley system could cause an asynchronous alternator's bearings to fail.

Connect the frame of the asynchronous alternator to earth ground.

If you are not certain that your wiring is correct have it checked by an electrician and have the electrician witness the testing of the asynchronous alternator.

The direction of rotation of the asynchronous alternator is important. The prime mover should tend to turn the asynchronous alternator in the same direction it would turn itself when it operates as a motor. If the asynchronous alternator has the wrong rotation it can be reversed. On some single-phase motors it is possible to reverse the starting winding connections. On other single-phase motors it is possible to transpose the motor's end bells so that the shaft comes out the other end.

*** CIRCUIT PROTECTION FOR THE ASYNCHRONOUS ALTERNATOR

Damaging overcurrents could occur if the prime mover overspeeds and by doing so sends too much power to the asynchronous alternator. Damaging overcurrents could also occur if the prime mover stalls or significantly slows.

An input fuse or circuit breaker should be used to protect the asynchronous alternator. The fuse or circuit breaker should not blow or trip when the asynchronous alternator is first connected to the electric utility's system. However, the fuse or circuit breaker should blow or trip if a current greater than rated nameplate current lasts for more than about 10 seconds.

If your electric service uses fuses, a time delay fuse should be used. Circuit breakers that are commonly used in modern electric services have a built-in time delay in their operation. The smallest time delay fuse or circuit breaker that allows the operation of the asynchronous alternator should be used. To estimate the size of your input fuse or circuit breaker, multiply the asynchronous alternator's nameplate current rating by 1.25. Select a time delay fuse that is equal to or less than the product of this multiplication.

*** POSSIBLE MODES OF OPERATION

Normal operation for an asynchronous alternator should be producing steady power that is less than the rated output of the asynchronous alternator. The prime mover must produce a near constant output to maintain this. If a waterwheel is the prime mover it should be receiving from a nearly constant water supply.

Overspeed operation occurs if the prime mover drives the asynchronous alternator faster than its maximum speed rating. This would occur with a waterwheel prime mover if the water supply level rose too high. During overspeed operation the asynchronous alternator produces more power and current than the alternator can handle. If an asynchronous alternator overspeeds for too long, it will overheat, destroy its electrical insulation, and fail.

Underspeed operation occurs if the prime mover does not produce enough power to rotate the asynchronous alternator faster than synchronous speed. This would occur with a waterwheel prime mover if the water supply level dropped too low. During underspeed operation the asynchronous alternator operates as a motor. It receives electric power from the electric utility and uses it to rotate the prime mover. If the prime mover is not easy to turn, the asynchronous alternator (now operating as a motor) could be overloaded. If it stayed in the overloaded condition for too long, it will overheat, destroy its electrical insulation, and fail.

Speed oscillations may occur if a prime mover is not steady. It may be serious enough to cause an electrical or mechanical overload. If there are oscillations, but there is no overload, the oscillations are probably not serious.

*** TESTING A NEWLY BUILT ASYNCHRONOUS ALTERNATOR

FIRST TEST: NO-LOAD TESTING

a) Start the prime mover connected to the asynchronous alternator, but with the asynchronous alternator disconnected from the utility's electric system.

b) Run the prime mover at the rated full-load speed of the asynchronous alternator. These speeds are shown on page 51. A tachometer or strobe will be necessary to measure the speed.

c) Allow the system to run unloaded for several hours. The alternator and prime mover should run quietly.

SECOND TEST: LOAD TESTING

   a) Start the system as in the no-load test.

   b) When the prime mover is up to speed, connect the output of the converter to the utility's electric system.

   c) Check the asynchronous alternator current, it should be slightly less than the rated nameplate current. If it is greater, quickly reduce the speed until the current drops. Overcurrent would cause the asynchronous alternator to overheat and burn through its electrical insulation.

   d) When the current is satisfactory measure the speed of the asynchronous alternator. The speed should be less than, but close to, the relevant maximum speed given in the table on page 49.

   e) Temporarily turn off all electrical devices connected to your home's electric system. This should cause your watthour meter to run backward. You should be able to see its disk turning in a reverse direction.

   f) Allow the asynchronous alternator to run loaded for several hours. During this time the asynchronous alternator will increase in temperature. Most modern induction motors can withstand temperatures up to 40°C greater than standard room temperature. This means a motor's case can go up to a maximum of about 140°F. If you can rest your hand on the asynchronous alternator it is not being overheated. If you can not rest your hand on it there may be a heating problem. If there is a burning insulation smell there is an overheating problem.

   g) If there is no overheating, that's great. You now have a functioning asynchronous alternator.

   h) If there is a heating problem, it will be necessary to reduce the power coming from the prime mover.

*** UNSTEADY OR INTERMITTENT PRIME MOVER

   It is possible to use an unsteady or intermittent prime mover like a windmill to drive an asynchronous alternator, but it is not easy.

   The following should be added to an asynchronous alternator system to make it satisfactory for operation with an unsteady prime mover.

First, there should be a speed switch that connects the asynchronous alternator to the electric utility system only when the prime mover's speed is above synchronous speed.

Second, there should be an underpower sensing circuit that detects when the asynchronous alternator is not sending enough power to the electric utility system and disconnects it.

Third, an overspeed or overpower sensing circuit should shut down the prime mover if too much power is being produced.

*** DC TO INVERTER METHOD OF PARALLEL POWER GENERATION

Another method of parallel power generation, without an asynchronous alternator, uses a DC voltage source to feed an inverter. Inverters change DC electric power to AC electric power electronically. The inverter creates AC voltage that can be connected to a utility's electric system.

This method shows some future promise with unsteady prime movers, like wind mills. A wind mill being driven by an unsteady wind could send varying power to a DC generator. The DC generator would then send the varying power to a battery bank. The battery bank would supply power to an inverter. Finally, the inverter would supply power to the electric utility. Presently, the equipment required to do this is too expensive. However, electronic prices are dropping. This method may be economically practical in the future.

*** REFERENCE

PURPA Handbook, published by the American Wind Energy Association, 122 C Street NW, 4th floor, Washington, DC 20001. This is a 14 page booklet that summarizes PURPA's legislative provisions. Order it from the American Wind Energy Association for $5.00/members and $7.50/nonmembers.

7. AC/DC POWER SUPPLIES

Power supplies used in electric power circuits are different from those typically used in electronic circuits. Power supplies for electronic circuits usually convert 120 volts AC down to a fixed 5 to 15 volts DC and supply no more than a few amperes at that low voltage. The power supplies for electric power circuits have outputs that are AC or DC, often higher than 15 volts, and capable of more than tens of amperes.

This chapter considers rugged inexpensive fixed and adjustable AC/DC power supplies that can produce up to 10 amps and 120 volts. These power supplies could be used to test motors or operate experimental equipment like the MHD propulsion system in chapter 9. Construction plans are provided for fixed transformer and adjustable autotransformer power supplies.

*** MAKING DC FROM AC WITH A BRIDGE RECTIFIER

The power supplies of this chapter all start with 60 Hz AC voltage. The conversion from AC to DC can be done many ways. It is done mechanically with devices such as the commutators on DC generators and electronically with many different vacuum tube and semiconductor devices.

Today the most common way of making DC from AC is the simple bridge rectifier circuit using silicon semiconductor diodes. This is shown in Figure 7-1.

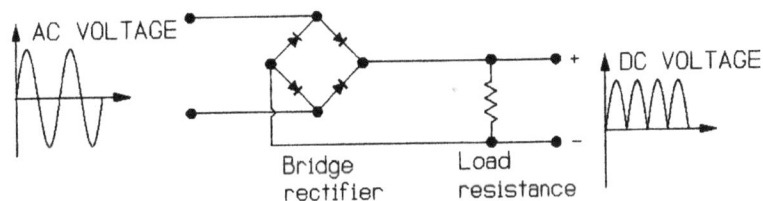

Figure 7-1 Diode bridge rectifier circuit connected to a resistance load. The input voltage is AC and the output voltage is full-wave rectified DC.

Often the output voltage of the diode bridge rectifier is smoothed by connecting a relatively large electrolytic filter capacitor across the rectifier circuit output. For a given resistance load, the larger the capacitance value the smoother the DC output voltage will be. Figure 7-2 shows this.

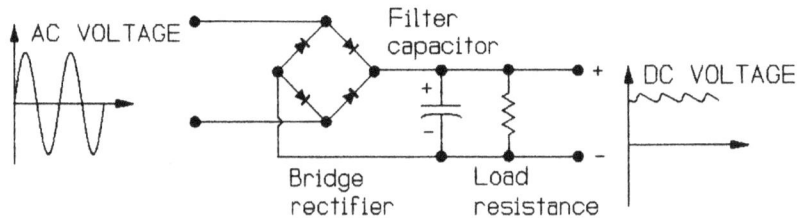

Figure 7-2 Diode bridge rectifier circuit with a filter capacitor connected to a load resistance. The input voltage is AC and the output voltage is filtered full-wave rectified DC.

There are several things that the experimenter should understand about electrolytic capacitors.

1) They can only handle one polarity of voltage, unless the applied voltage is very low. If an electrolytic capacitor is connected with the wrong polarity to a voltage close to its rated value, it will conduct current like a resistor. This will cause it to overheat and possibly explode from overheating. The power of the explosion depends on the voltage, capacitor design, capacitance, and other variables. Roughly speaking an electrolytic capacitor about 3/8" in diameter and 1" long is capable of the explosiveness of a small firecracker. A larger capacitor would have more explosive energy.

2) The voltage rating of an electrolytic capacitor should be heeded. It could explode, if the voltage applied to it is beyond its rating.

3) Because of the explosive capability of electrolytic capacitors, it is usually best to put a cover over them.

4) Electrolytic capacitors can store a significant amount of charge for a long time. To decrease the amount of time that they store charge, it is a good idea to have discharge resistors in parallel with larger electrolytic capacitors.

*** TYPES OF POWER SUPPLIES

A power supply is a unit that supplies electrical power to another unit. The input voltage to a power supply can be any voltage. However, for the supplies of this chapter the input voltage is 120 volts AC at 60 Hz.

The simplest way of producing a voltage that is less than the supply voltage is the voltage divider circuit. This circuit can use two fixed resistors or one variable potentiometer. It is shown in Figure 7-3. The voltage divider circuit will work with electrical power circuits, but it loses a great deal of power as heat. Sometimes 100 Watt incandescent light bulbs can be used as the resistors in a voltage divider circuit to provide a useful voltage to a temporary circuit. The AC output of a voltage divider can be easily converted to DC with the circuits of Figures 7-1 and 7-2.

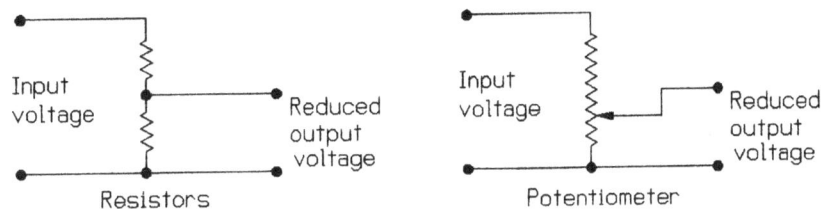

Figure 7-3 Voltage divider circuits.

Modern light dimmers can be practical sources of variable AC voltage for incandescent light bulbs and other resistive loads. The heart of the light dimmer is a TRIAC, an electronic device that only conducts when it has received a sufficient voltage on its GATE terminal. The TRIAC will conduct during all of an AC cycle, just part of it or none of it depending on when, and if, its GATE terminal receives voltage during an AC cycle. An operating TRIAC turns on suddenly twice in every AC cycle. This sudden turn on creates high frequency EMI that may be picked up by sensitive electronic circuits. Light dimmers are inexpensive and can be purchased at hardware stores. The AC output of a light dimmer can be converted to DC with the circuits of Figures 7-1 and 7-2.

Transformers are often used to convert one AC voltage to another. They were discussed on pages 11 to 13. Transformers are rugged, efficient, and easily available from scrapped surplus equipment and electronics parts stores. The AC output of a transformer is often converted to DC with the circuits of Figures 7-1 and 7-2.

Variable autotransformers (commercially called VARIACs or POWERSTATs) are used to convert one AC input voltage to a variable output AC voltage. A variable autotransformer has one winding wound around a donut shaped laminated steel core. Input AC voltage is applied to the two terminals of the winding. The winding is specially made so that bare sections of its wires make electrical contact with the third terminal, a moveable conducting carbon brush. Output AC voltage is taken from one of the winding terminals and the moveable third terminal. Variable autotransformers are expensive. A new one rated for 120 volts AC input and 0 to 140 volts AC output at 10 amps may cost $150.00. Of course, scrapped or surplus variable autotransformers can be much less expensive. Schematics and photographs of variable autotransformers are shown in Figures 7-7 and 7-8. The AC output of a variable autotransformer is easily converted to DC with the circuits of Figures 7-1 and 7-2.

Batteries can be used as a simple DC power supply. Automobile batteries can produce high currents. However, they are not adjustable and are heavy to move around. Flashlight cells (batteries) have too low a voltage and current capacity by themselves. Connected in series and parallel combinations they could be used in electric power circuits, but generally too many of them would be required.

An automobile battery charger is another simple DC power supply that can be used in some electric power circuits. However, the output is fixed (although some chargers have a switchable output, one for 6 volt batteries and one for 12 volt batteries).

*** CONSTRUCTION OF A FIXED TRANSFORMER SUPPLY SYSTEM

BILL OF MATERIALS

ITEM	QUANTITY	WHERE OBTAINED
Filament transformer 115/12.6 volts AC 60 Hz with a 20 amps rating on the 12.6 volts AC output	One	Scrapped Equipment
AC to DC bridge rectifier, with at least a 50 volts DC and 20 amps DC ability	One	Electronics parts store
Heat sink	One	Scrapped equipment or copper strips (I used copper water pipe that was hammered flat and formed into a heat sink)
15000 μF or greater electrolytic capacitor with at least a 35 volts DC rating	One	Electronics parts store or scrapped equipment

3300 ohm ½ watt resistor	One	Electronics parts store
20 amps fuse and fuse holder	One	Electronics parts store
Power cord and outlet plug	About 6 feet long	Electronic parts store or scrap wire
¼" x 1 ½" bolts for terminals	Four	Hardware store
¼" nuts to fit on the terminal bolts	Twelve	Hardware store
Flat washers to fit on the terminal bolts	Twenty	Hardware store
Solder	Short piece	Hardware store
Black plastic electrical insulating tape	Less than 1 roll	Hardware store
Assorted colors of #14 insulated copper wire	Several feet	Scrapped Romex wire
Flat headed wood screws for mounting the transformer, capacitor, fuse holder, and bridge rectifier	Eight	Hardware store
Large staples for holding the power cord to the mounting board	Two	Hardware store
Wooden mounting board 9 ¼" x 12" x ¾"	One	Lumber yard or scrap
Plywood for the cover box	About 3 sq. ft.	Lumber yard or scrap
Molding about ¾" x ¾" for the cover box	About 50 inches	Lumber yard of scrap
Small nails for the cover box	About 50	Hardware store
Wood glue for the cover box	Small amount	Hardware store

CONSTRUCTION PROCEDURE:

1) Study Figures 7-4 to 7-6.

Figure 7-4 Photograph of the fixed transformer power supply. Notice that there are two transformers in this photograph. This supply had two parallel transformers to increase the output capability. Also notice the wooden box behind of the transformers. The box fits over the supply when the supply is operating. It is designed to contain the filter capacitor in the unlikely event that it explodes.

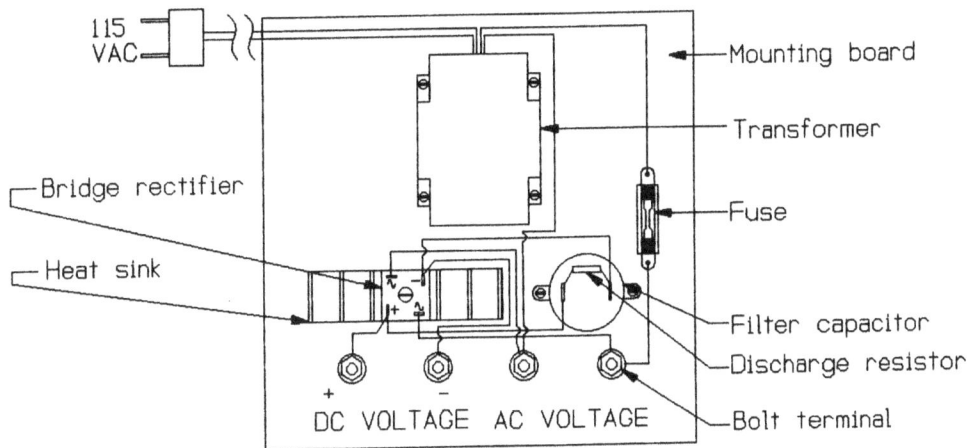

Figure 7-5 Physical drawing of the fixed transformer power supply.

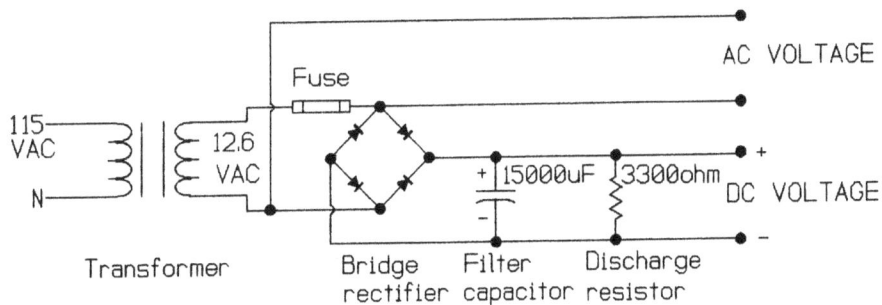

Figure 7-6 Schematic drawing of the fixed transformer power supply.

2) Have the following tools available.

Electric drill, assorted bits, and a center punch
Wood saw
File
Sandpaper
Screw driver
Soldering gun
Pencil
Carpenter's square

3) Gather the needed parts and materials.

4) Connect the input leads of the transformer to the power cord. Solder the connections and cover them with electrical tape.

5) Arrange the transformer, filter capacitor, bridge rectifier with its heat sink, terminal bolts, and fuse holder on the mounting board. Leave enough space between the parts for the connecting wires. Mark the locations.

6) Drill and countersink the mounting board for the terminal bolts. The heads of the terminal bolts should fit into countersunk holes in the bottom of the mounting board so that they will not protrude beyond the bottom surface of the mounting board.

7) Mount the parts on the mounting board. Use large staples to hold the power cord to the board. Cover the bolt heads on the bottom of the mounting board with a piece of plastic electrical tape. This will electrically insulate them if the power supply is set on a conducting surface.

8) Connect the circuit components using #14 AWG copper wire. Be neat. Solder the wires to each terminal and to each other as indicated by the schematic diagram in Figure 7-6. Use plastic electrical tape to cover the wire to wire solder connections. Visually double check that the circuit agrees with the schematic diagram and that there are no accidental dripped solder shorts. Be especially certain that the filter capacitor is connected with the correct polarity connections to the bridge rectifier.

9) Put a fuse into the fuse holder that is rated to the output current capacity of the transformer or the current capacity of the bridge rectifier, whichever is less.

10) Build a wooden box cover for protection against the unlikely possibility of a filter capacitor explosion. Put holes in the rear of it and air space at the front bottom to allow for cooling air circulation. The box I built is shown in Figure 7-4.

TEST PROCEDURE:

Measure the resistance from prong to prong on the power cord plug with an ohmmeter. The resistance should not be zero. If it is zero there is a short circuit. In that case, recheck the circuit and correct it. The resistance value will depend on the capacity of the transformer. On my power supply the resistance is 6.9 ohms. Note the 6.9 ohms is DC impedance. Effective AC impedance is higher, but can't be measured directly with an ohmmeter.

Plug in the power cord. Measure the voltages. The AC VOLTAGE terminals should have about 12.6 volts AC and 0 volts DC across them. The DC VOLTAGE terminals should have 1.41 times the AC voltage measured across the AC VOLTAGE terminals. This is about 12.6 x 1.41 = 17.8 volts DC. The AC voltage measured across the DC VOLTAGE terminals should be very small or zero.

*** CONSTRUCTION OF A VARIABLE AUTOTRANSFORMER SUPPLY SYSTEM

BILL OF MATERIALS

ITEM	QUANTITY	WHERE OBTAINED
Variable autotransformer 120 to 0-140 volts AC with at least a 7.5 amps output and mounting hardware	One	Scrapped equipment
AC to DC bridge rectifier, with at least a 200 volts DC 7.5 amps DC rating	One	Electronics parts store
Heat sink	One	Scrapped equipment or copper strips (I used copper water pipe that was hammered flat and formed into a heat sink)
7 ½ amps fuse and fuse holder	One	Electronics parts store
Power cord and outlet plug	About 6 feet long	Electronics parts store or scrap wire
Terminal posts	Four	Electronics parts store
Solder	Short piece	Hardware store
Assorted colors of #14 insulated copper wire	Several feet	Scrapped Romex wire
Flat headed wood screws for mounting the bridge rectifier and the fuse holder	Two	Hardware store
Large staples for holding the power cord to the mounting board	Two	Hardware store
Plywood mounting boards	About three square feet	Lumber yard or scrap
Small nails	About 20	Hardware store
Wood glue	Small amount	Hardware store

CONSTRUCTION PROCEDURE:

1) Study Figures 7-7 to 7-9.

Figure 7-7 Photograph of the two variable autotransformer power supplies.

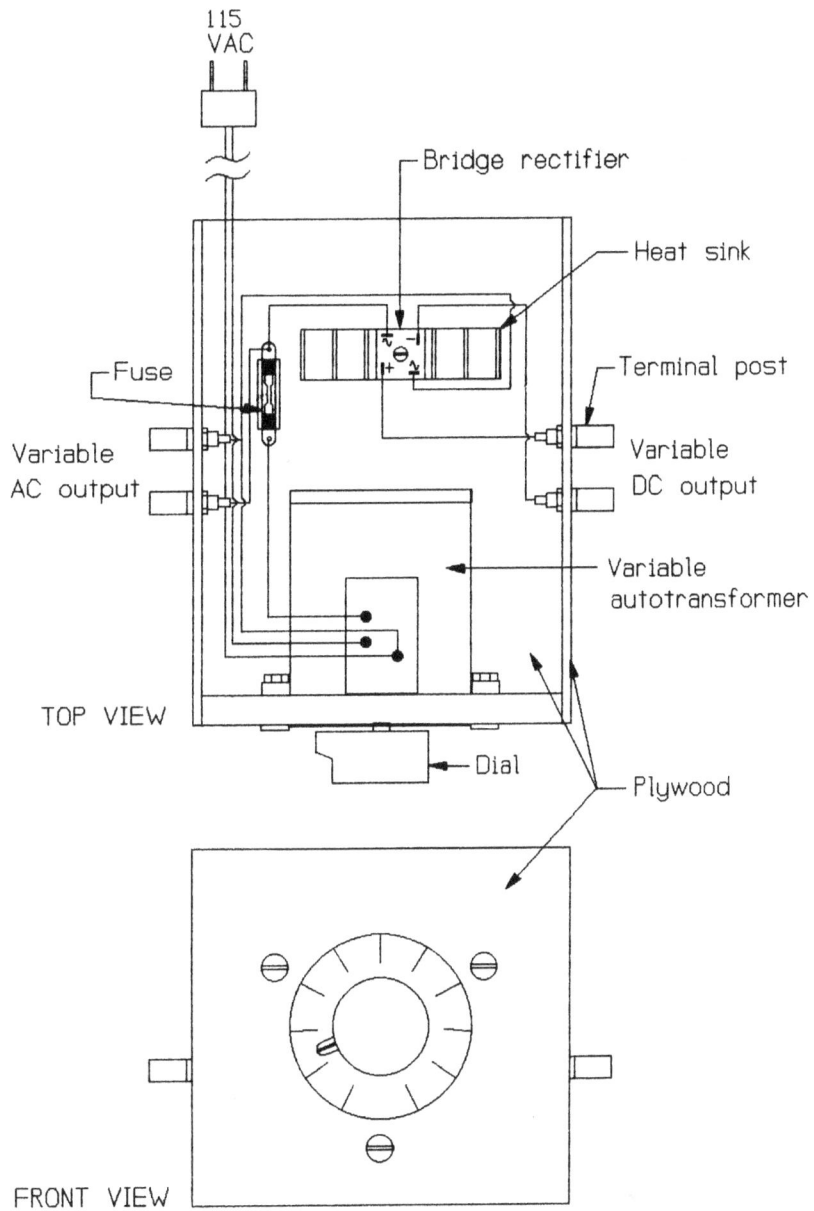

Figure 7-8 Physical drawing of a variable autotransformer power supply.

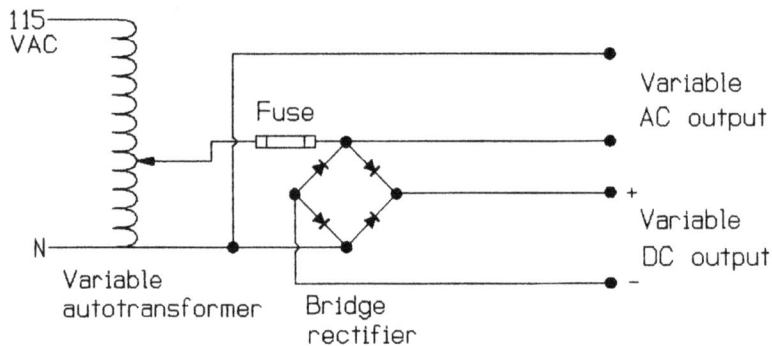

Figure 7-9 Schematic drawing of the variable autotransformer power supply.

2) Have the following tools available.

        Electric drill, assorted bits, and a center punch
        Wood saw
        File
        Sandpaper
        Screw driver
        Soldering gun
        Pencil
        Carpenter's square

3) Gather the needed parts and materials.

4) Build a four sided open mounting box from plywood, like the one shown in Figures 7-7 and 7-8. The box I built was 8 1/2" across by 8" high by 10 3/4" deep. The box is open rather than closed to allow better cooling air circulation and to make it easier to replace fuses. However, the open box does increase the possibility of electric shock. If you feel there is a danger that someone, or something, may touch an uncovered live terminal or the winding of the variable autotransformer, then an enclosing cover should be built. If an enclosing cover is built, be certain to leave holes at the top and bottom for air circulation.

5) Arrange the variable autotransformer, bridge rectifier with its heat sink, terminals, and fuse holder on the mounting box. Leave enough space between the parts for the connecting wires. Mark the locations.

6) Mount the parts on the open mounting box. Use large staples to hold the power cord to the board. Do not puncture the cord with the staples.

7) Connect the circuit components using #14 AWG copper wire. Be neat. Connect and solder the wires as indicated in the schematic diagram of Figure 7-9. Visually double check that the circuit agrees with the schematic diagram and that there are no accidental dripped solder shorts.

8) Put a fuse into the fuse holder that is rated to the output current capacity of the variable autotransformer or the current capacity of the bridge rectifier, whichever is less.

TEST PROCEDURE:

Measure the resistance from prong to prong on the power cord plug with an ohmmeter. The resistance should not be zero. If it is zero there is a short circuit. In that case, recheck the circuit and correct it. The resistance value will depend on the capacity of the variable autotransformer. On my power supply the resistance is 2.1 ohms. Note the 2.1 ohms is DC impedance. The effective AC impedance is higher, but can't be measured directly with an ohmmeter.

Plug in the power cord. Measure the voltages. When the input voltage is 120 volts AC, the AC output voltage should be 0 to 140 volts AC, depending on the dial setting. Many variable autotransformers have several connection taps. If the lower taps were used, the output might be between 0 and 120 volts AC. The AC output terminals should have 0 volts DC across them. The DC output terminals should have 0 to 140 volts DC across them. The AC voltage measured across the DC VOLTAGE terminals should be between 0 and 16 volts AC.

*** MANUFACTURER

Superior Electric
383 Middle Street
Bristol, Ct 06010
860-585-4500
http://www..superiorelectric.com

They call their variable autotransformers "Powerstats". These are made for a variety of AC frequencies and power capabilities, for single-phase or three-phase, and are motor-driven or manually controlled. All are described in their free catalog. I used "Powerstats" in the variable autotransformer power supplies that I described in the above plans.

8. THERMOELECTRIC GENERATOR

A thermoelectric generator uses the thermoelectric effect (also called Seebeck effect) to produce DC electricity from heat. It uses a temperature difference between a hot source and cold sink (cold source). The greater the temperature difference is between the hot source and the cold sink, the greater is the magnitude of generated DC voltage. If the thermoelectric generator's terminals are connected to a resistive load, it will produce electrical power.

There are two major types of thermoelectric generators, metal to metal thermocouples and semiconductor thermoelectric modules.

The most common thermoelectric generators are metal to metal thermocouples. These use junctions of dissimilar metals. There are one or two reference junctions and one measurement junction. Figure 8-1 shows an iron/constantan, type J, thermocouple being used to measure the temperature of hot water.

To accurately measure temperature, the temperature of the reference junctions must be known. Sometimes the reference junctions are placed in ice water baths to fix their temperatures at 32°F. This is shown in Figure 8-1. Other times the reference junctions are simply exposed to room temperature air. Then the room temperature must be known to accurately determine temperature from the thermocouple's output voltage.

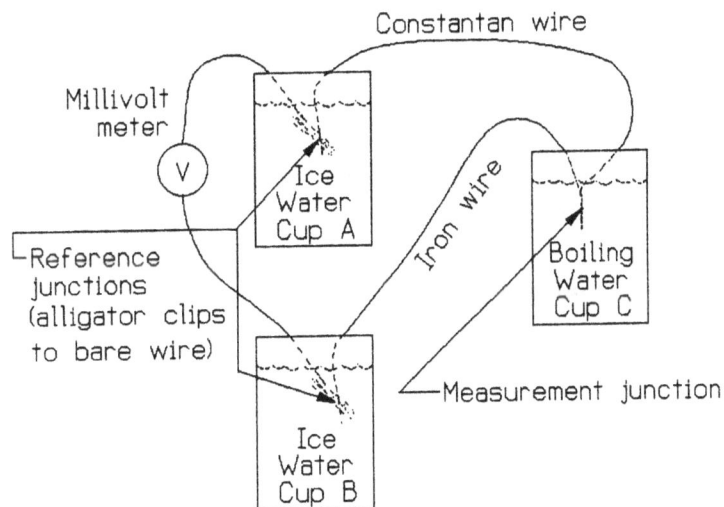

Figure 8-1 Two-reference junction thermocouple circuit being used to measure the temperature of a cup of boiling water.

It is possible to use just one reference junction, if one thermocouple wire is made of the same metal as is used in all meter leads and meter wiring. This is shown in Figure 8-2 where one thermocouple wire is copper in a copper/constantan, type J, thermocouple wire pair.

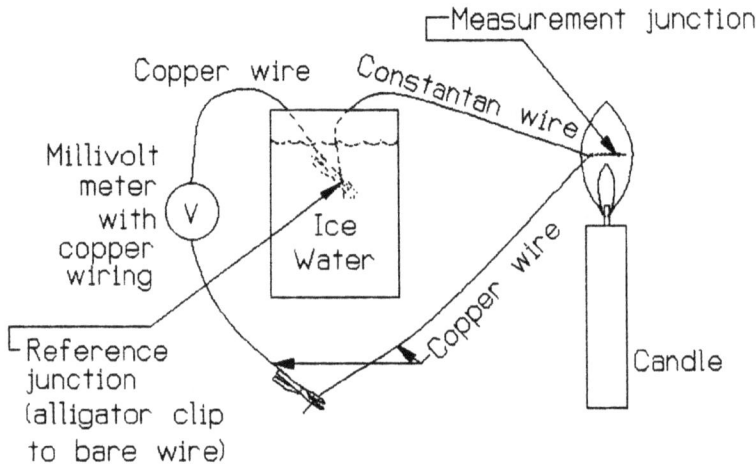

Figure 8-2 One-reference junction thermocouple circuit being used to measure the temperature of a candle flame.

There are many metal combinations used in thermocouples. It is only required that two different metals are used in a thermocouple. Some standard metal pairs used in industry are: chromel/constantan (type E), iron/constantan (type J), chromel/alumel (type K), and copper/constantan (type T).

Semiconductor thermoelectric generator modules use metal to semiconductor junctions. Like metal to metal thermocouples they require a hot source and cold sink. Unlike metal to metal thermocouples, they can produce a significant amount of electrical power. A semiconductor thermoelectric generator module uses two types of semiconductor material, P type material and N type material. The P and N type semiconductor materials are doped (mixed with precise amounts of chemical impurities) in the same way semiconductors are doped in diodes and transistors. The semiconductor materials are sliced into pellets and joined to metal terminals at two ends. Figure 8-3 shows a two-pellet semiconductor thermoelectric generator module producing electrical power.

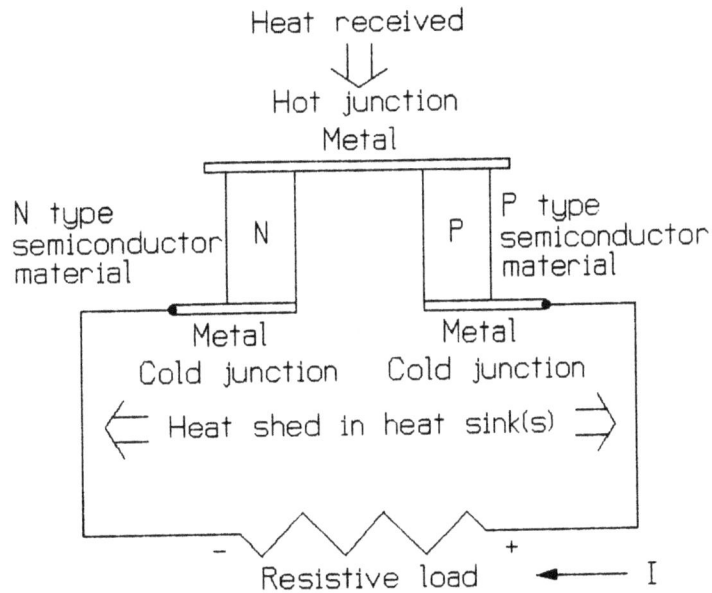

Figure 8-3 Simplified two-pellet semiconductor thermoelectric generator module.

Usually semiconductor thermoelectric generator modules have more than two pellets. Figure 8-4 shows a typical, many pellet, module.

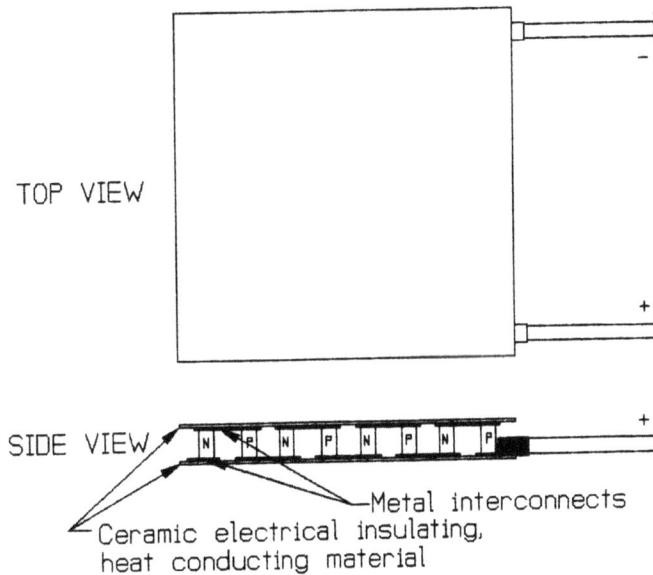

Figure 8-4 Typical commercially available thermoelectric generator module.

*** WHERE TO USE A THERMOELECTRIC GENERATOR?

Thermocouples can only produce small quantities of electric power. However, they are very useful in temperature measurement and are often used to measure extreme high or low temperatures. Thermocouples are inexpensive, rugged, and accurate for temperature measurement. One example of their use is in power plants where thermocouples measure boiler and steam temperatures. Other examples of their use are in the safety systems of a home's gas hot water heater and furnace. On these, gas pilots must be lit to allow the main gas valves to open. Thermocouples are positioned over the gas pilots and then connected to electromagnetic coils in the gas valves. When a thermocouple sends sufficient current to its electromagnetic coil a latch opens, thus allowing the main gas valve to open when the thermostat switch requests it.

Semiconductor thermoelectric generator modules can produce significant amounts of electrical power. They are useful in locations requiring reliable electric power for longer periods of time than batteries or rotating electric generators can provide. Kerosene powered thermoelectric converters are used in the countries of the former Soviet Union to provide power for radio receivers in remote areas. Thermoelectric generators powered by small nuclear reactors are used on U.S. space probes.

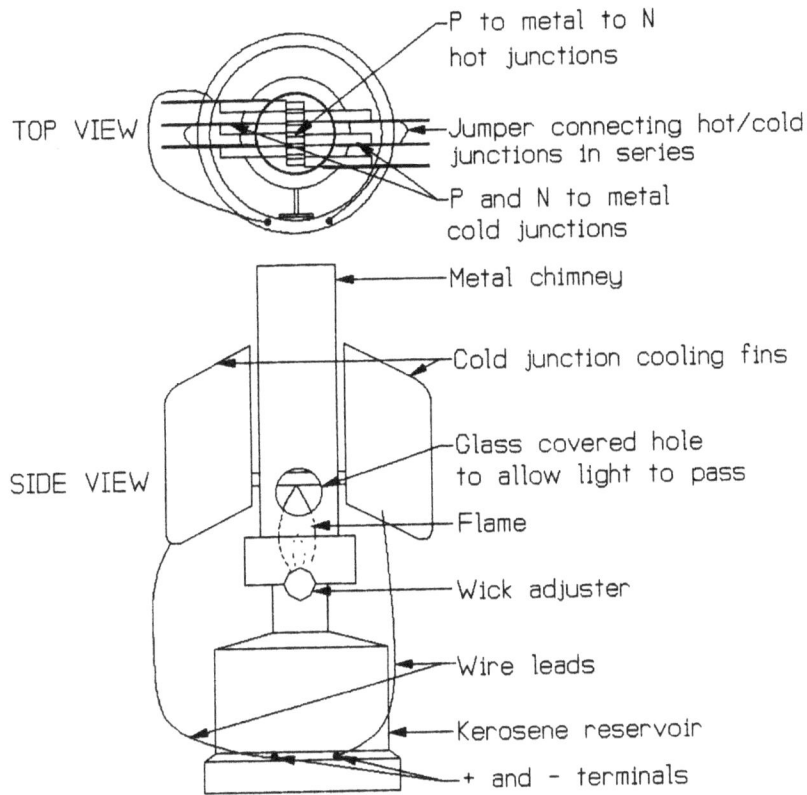

TOP VIEW

- P to metal to N hot junctions
- Jumper connecting hot/cold junctions in series
- P and N to metal cold junctions

SIDE VIEW

- Metal chimney
- Cold junction cooling fins
- Glass covered hole to allow light to pass
- Flame
- Wick adjuster
- Wire leads
- Kerosene reservoir
- + and - terminals

Figure 8-5 Simplified sketch of a kerosene powered thermoelectric generator. Units like this can produce about .2 watts at 9 volts.

*** WHAT IS THERMOELECTRIC COOLING?

Most semiconductor thermoelectric modules are not used as generators at all, but as heat pumps for small refrigerators. By applying sufficient DC voltage the modules will pump heat from the center junctions to the outer junctions. This is the same as cooling the center junction. The effect is now being used in special picnic basket "ice" chests that you can power from your car's battery and in many other refrigeration applications. The following figure shows the cooling effect with a two-pellet module.

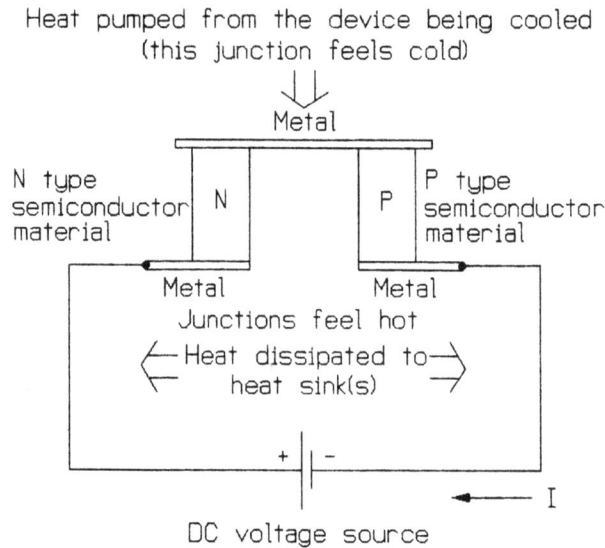

Figure 8-6 Simplified sketch of a two-pellet semiconductor refrigeration module.

*** WHO SHOULD USE A THERMOELECTRIC GENERATOR?

Thermocouples are of interest to the experimenter. Different combinations of metals will produce different voltages for the same junction temperature differences. Different parallel and series circuit combinations and conductor sizes will produce different voltages and currents.

Thermocouples are of use in measuring temperatures beyond the temperature range of a thermometer or in locations where a thermometer can not be placed. Thermocouples can be used to measure the temperature of dry ice, hot gas, steam or flames. Thermocouples can be imbedded in equipment for temperature monitoring. For example, thermocouples can be placed in the windings of an electric motor to determine internal temperatures during motor operation.

Semiconductor thermoelectric generators are of interest to those wishing a long term low wattage silent and continuous DC electric power source where batteries won't last and solar cells won't work. Anybody who wants to run electrical equipment for several months in a cave or in northern Siberia in the winter could use a kerosene flame powered semiconductor thermoelectric generator.

Semiconductor thermoelectric generators are of interest to the experimenter. It may be possible to improve on the present semiconductor thermoelectric generators with different materials. If low cost semiconductor thermoelectric generators could be made, perhaps the temperature difference between a hot solar collector and cool well water would be enough to be economically generate electric power.

*** WHO SELLS THERMOELECTRIC GENERATORS?

THERMOCOUPLES

In a large city, you can often buy thermocouple wire and metering equipment directly from a dealer. Here, in Pittsburgh, PA, there are three vendors listed under thermocouples in the *Business to Business* telephone directory.

Thermocouples can be purchased through the mail. It is worthwhile to contact some larger companies that specialize in thermocouples and thermocouple products, even if you don't purchase their products. The free catalogs they provide often contain a great deal of technical information on thermocouples. Two of the companies are:

Thermodynamic Sensors
1193 McDermott Dr.
West Chester, PA 19380
1-800-523-2002
http://www.thermodynmicsensors.com

Tudor Technology, Ltd.
5145 Campus Drive
Plymouth Meeting, PA 19462
1-800-777-0778

SEMICONDUCTOR THERMOELECTRIC GENERATORS

The following companies will sell small quantities of thermoelectric cooling modules. Their modules are designed for cooling, but will operate as electric generators. As with the thermocouple companies, it is worthwhile contacting them and requesting their free catalogs. International Thermoelectric, Inc. in their catalog no. 100 even provides a detailed design example for a semiconductor thermoelectric generator.

International Thermoelectric, Inc.
131 Stedman Street
Chelmsford, MA 01824
1-508-452-0212
fax 1-508-452-0212

Tellurex Corporation
1248 Hastings St.
Traverse City, MI 49684
1-231-947-0110
http://www.tellurex.com

*** WHAT IS THE BASIC PRINCIPLE OF THE THERMOELECTRIC EFFECT?

Thomas Seebeck discovered the thermoelectric effect in 1821. He incorrectly believed that the effects he produced with his thermocouples and semiconductor thermoelectric generators were caused by the flow of heat. Electricity was not well understood then. Seebeck did not realize he was producing electricity.

Now it is believed that the thermoelectric effect produces electricity by causing the outer shell electrons of one material's atoms to be more energetic than the other material's. This makes the electrons more likely to move away from one material than the other, thereby causing an electrical polarity difference between the materials.

Modern explanations of the thermoelectric effect use quantum mechanics. The explanations consider electrons, valence bands, holes, and Fermi levels. The explanations become very confusing, very quickly.

*** THERMOCOUPLE DEMONSTRATION EXPERIMENT #1

OBJECT

To show the operation and use of a two-reference junction thermocouple system.

MATERIALS

     One 12 inch length of an iron/constantan, type J, thermocouple wire pair
     One millivoltmeter (Electronics parts stores have these)
     Two small alligator clip leads (Electronics parts stores have these)
     Three coffee cups
     Ice water
     Boiling water
     Two Pliers (needle nose are best)

PROCEDURE

1) Carefully clean the ends of all the wires and scrape them with a knife to expose shiny metal.

2) Tightly twist one end of the constantan wire to the end of the iron wire. Make this twisted junction 3/16" or less in length. This is where having two pliers is useful

3) Bend the wires so that the twisted together portion sits in one empty coffee cup and the other two ends each go into an empty coffee cup. The cup arrangement is shown in Figure 8-1.

4) Using the alligator clip leads, connect the millivolt meter to the thermocouple wires as shown in Figure 8-1.

5) Turn on the meter with it set to DC millivolts. The meter should read 0 millivolts. If the meter does not read 0 then record the amount it does read, that is the zero offset error. In future readings subtract the zero offset error to get the actual reading.

6) Put ice cubes and ice water in cups A and B. Leave the measurement junction in the air.

7) Note the meter reading. If the room temperature is 70°F (20.9°C), the reading should be about 1.1 millivolts.

8) Pour boiling water into cup C. The temperature of boiling water would produce a thermocouple voltage of about 5.3 millivolts. However, the water in the cup soon cools to less than boiling so the readings will not be quite as high as 5.3 millivolts.

*** THERMOCOUPLE DEMONSTRATION EXPERIMENT #2

OBJECT

To measure the temperature of a candle flame at various locations.

MATERIALS

The same as in experiment #1, except that the cup of boiling water is replaced with a lit candle.

PROCEDURE

1) Use the circuit of experiment #1.

2) Hold the measurement junction at different locations in the candle flame. At each location after a short time the voltage measurement should stabilize and indicate a constant voltage on the meter. It would also be interesting to dip the measurement junction into the liquid candle wax beneath the wick.

3) Use the thermocouple output voltage versus temperature graph of Figure 8-7 to determine the temperature at each location.

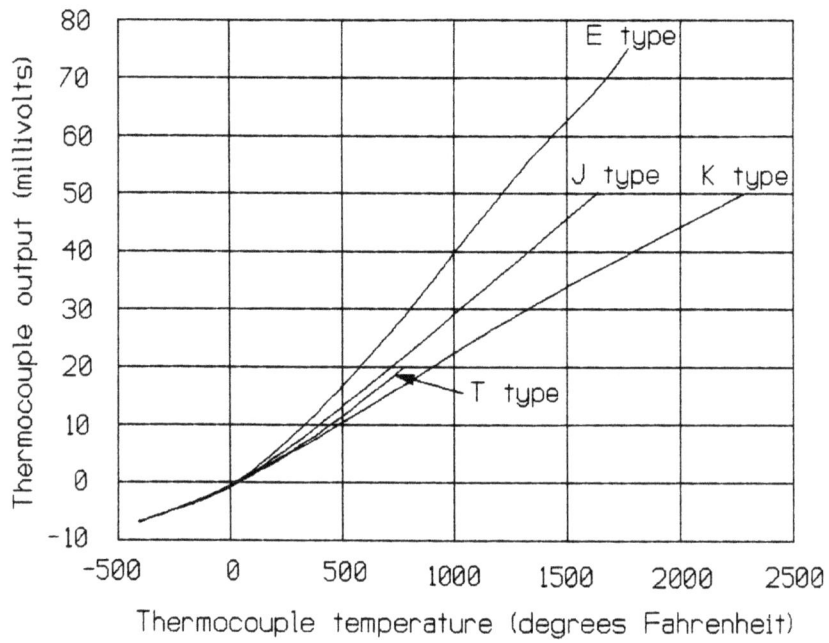

Figure 8-7 Thermocouple temperature versus voltage. The reference junctions are at 32°F (0°C).

9. MAGNETOHYDRODYNAMIC (MHD) PROPULSION SYSTEM

Magnetohydrodynamics is the study of phenomena arising from the motion of electrically conducting fluids in the presence of electrical and magnetic fields. Magnetohydrodynamic effects are of concern to those studying the earth's geomagnetic field, plasmas in thermonuclear fusion, and other areas. Magnetohydrodynamic effects have been used in experimental power generation systems and sea going (salt water) ship propulsion systems.

*** HOW IS MAGNETOHYDRODYNAMIC (MHD) PROPULSION FITTED TO A SHIP?

MHD propulsion requires large quantities of DC electrical power. A ship using MHD propulsion would have a DC generating plant on board. Where the propellers would ordinarily be, immersed in sea water, the MHD electrodes and electromagnetic field coils are attached. Figure 9-1 shows a surface ship fitted for MHD propulsion.

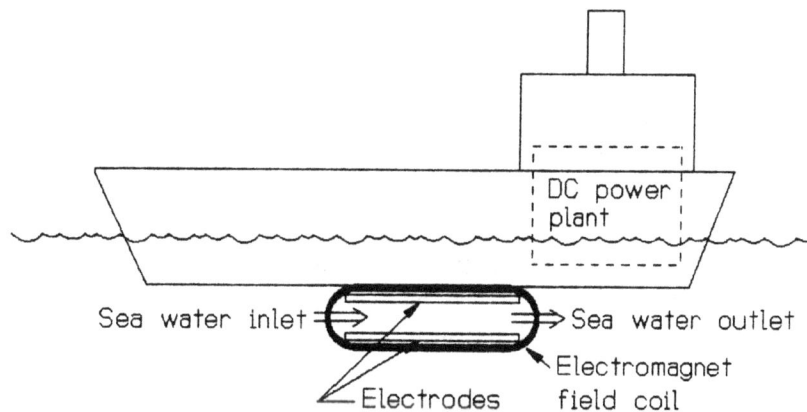

Figure 9-1 Simplified sketch of a sea going (salt water) ship fitted for MHD propulsion.

The circuit diagram for an MHD propulsion system includes two circuits, the electromagnetic field coil circuit and the electrode circuit. Figure 9-2 shows the schematic for each of these.

Figure 9-2 MHD propulsion system basic power schematic.

*** HOW DOES MAGNETOHYDRODYNAMIC (MHD) PROPULSION WORK?

Magnetohydrodynamic (MHD) propulsion is produced by the interaction of ionic electric current with a magnetic field.

Ionic electric conduction is required for MHD propulsion. Ionic conduction uses positive and negative ions as the charge carriers. Positive ions are atoms or groups of atoms that are missing one or more electrons. Negative ions are atoms or groups of atoms that have one or more extra electrons. Ionic conduction is different from the electric current that occurs in conductors where the atoms remain stationary and free moving electrons are the charge carriers.

The salt in sea water is a necessary part of an MHD propulsion system. The salt improves the conductivity of water by providing ions, mostly sodium and chlorine. Without salt the conductivity of water is very low. An MHD propulsion system would not work well in fresh water.

The immersed portion of an MHD propulsion system has two major components. First, it has positive and negative polarity electrodes. Second, is has strong electromagnetic field coils that produce magnetic flux at a right angle to the gap between the electrodes. Figure 9-3 shows this.

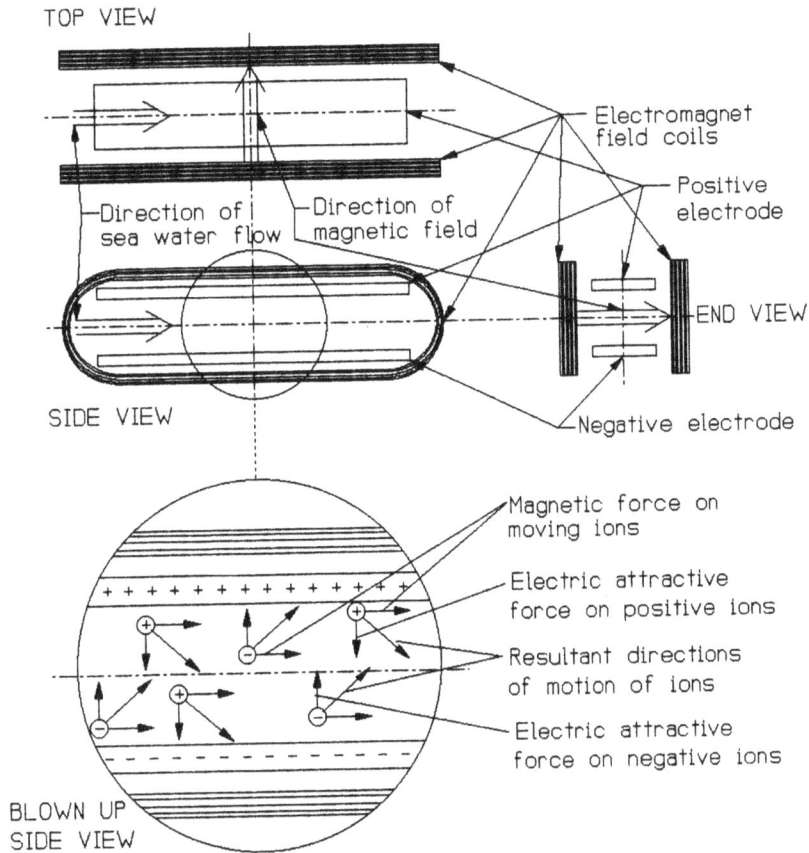

Figure 9-3 MHD propulsion system electromagnetic field coils and electrodes.

The electrodes are immersed in sea water. DC voltage causes a current to flow from one electrode to the other. The positive polarity electrode receives electrons from negative ions that are in the sea water. The negative polarity electrode donates electrons to positive ions that are in the sea water. The ions are attracted to their respective electrodes by electrical attraction.

In the MHD propulsion system while the ions move toward the opposite electrode they are also pushed toward the sea water outlet. This is shown in the BLOWN UP SIDE VIEW in Figure 9-3. The ions are pushed toward the sea water outlet because the magnetic field interacts with the motion of the electrically charged ions. If an ion happens to bump into a neutral atom or molecule, it bumps it in the same direction, toward the outlet. The result is a flow of ions, neutral atoms, and molecules toward the outlet. This flow is the propulsion push.

The reason that a moving charged particle going through a magnetic field is pushed is not known. It is a basic phenomenon of electromagnetism. A charged particle moving at a right angle to a magnetic field will receive a force at a right angle to the motion of the charged particle and at a right angle to the field. The direction of force is determined by Fleming's right and left hand rules. These rules are shown in Figure 9-4.

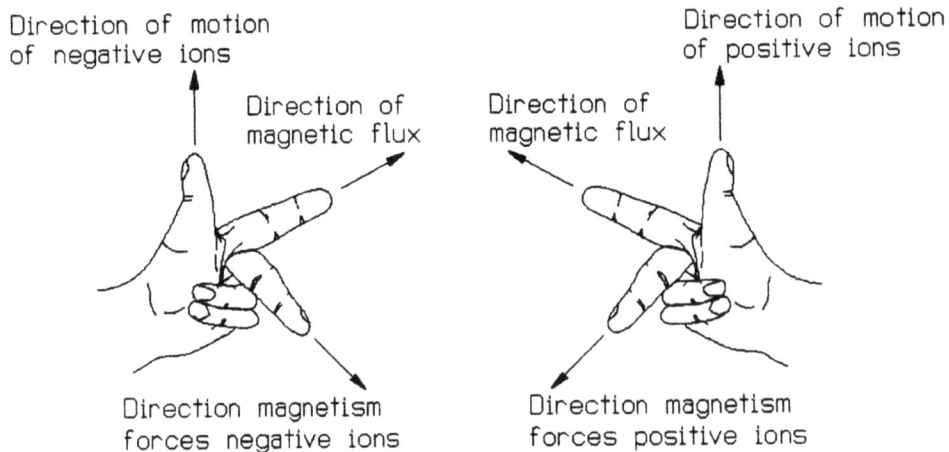

Figure 9-4 Fleming's right and left hand rules. (Note: the direction of magnetic flux goes from the North magnetic pole to the South magnetic pole.)

*** ENERGY LOSSES DURING MHD PROPULSION

Electrolysis will occur during the operation of an MHD propulsion system. Electrodes operating in sea water will liberate hydrogen and chlorine gas. The greater the current between the electrodes, the greater the amount of liberated hydrogen and chlorine there will be. The gases are not easily recoverable at sea, so the energy used in creating them is wasted.

The equation for the gas production is:

$$2NaCl + 2H_2O + electrolysis => Cl_2 + H_2 + 2NaOH$$

Another problem is the electrical erosion of the positive electrode (anode). This is the same action that occurs with the sacrificial donor anode during electroplating. The positive electrode is gradually dissolved into the salt water. The energy that does this does not contribute to propulsion, it is wasted.

A third problem is simple resistive heating. Energy is lost to heating the salt water while electrical current goes through it. This is also wasted energy.

*** SUPERCONDUCTING FIELD COILS

The stronger the MHD system's electromagnetic field, the more efficient the system will be. Electromagnetic field coils using superconductors can produce very strong magnetic fields without resistive heat losses in their conductors. For that reason the larger research projects on MHD propulsion have used superconducting coils.

So far the large research projects using superconducting coils have used the low temperature superconductors, the type cooled with liquid helium. The newer, higher temperature, liquid nitrogen superconductors are still not practical for producing fields of the strengths needed in a MHD system.

However, liquid helium is far too expensive for the experimenter on a budget. Because of the expense and difficulties of using liquid helium, superconducting field coils are not used in the experimental MHD project plans presented here.

*** CONSTRUCTION OF AN EXPERIMENTAL MHD PROPULSION SYSTEM

BILL OF MATERIALS

ITEM	QUANTITY	WHERE OBTAINED
Steel ring	One	Scrap yard
#22 AWG copper magnet wire	520 ft. (1 lb.)	Electric motor repair shop
#14 AWG plastic insulated copper wire	1 ft.	Left over house wiring
Casserole dish	2 quart	Kitchen
Table salt	7 teaspoons	Kitchen
Fresh water	1.5 quarts	Kitchen faucet
Solder and plumber's solder flux	Short piece of solder, dab of solder flux	Hardware store
Black plastic electrical insulating tape	2 rolls	Hardware store
Enamel spray paint	One can	Hardware store
Alligator clip leads	Eight	Electronics parts store

0 to 10 amp DC ammeter (a multimeter with scales in this range will do)	Two*	Electronics parts store
0 to 100 volts DC voltmeter (a multimeter with scales in this range will do)	Two*	Electronics parts store
Ohmmeter (a multimeter will do)	One	Electronics parts store
0 to 60 volts DC, 6 amps variable power supply	Two	See section titled DC POWER SUPPLIES

* It is possible to operate the MHD system with fewer meters. However, testing will be more awkward.

CONSTRUCTION PROCEDURE:

1) Find or make a magnetic steel or iron ring of the approximate dimensions shown in Figure 9-5. Use a permanent magnet to check that the material of your ring can be magnetized. The cross sectional area of the ring I used is 1"x(4"-2.75") = 1.25 square inches. The cross sectional area of the ring you use should be at least .8 square inches.

2) Have the following tools available.

> Hacksaw
> File
> Sandpaper
> Soldering gun
> Pencil
> Carpenter's square

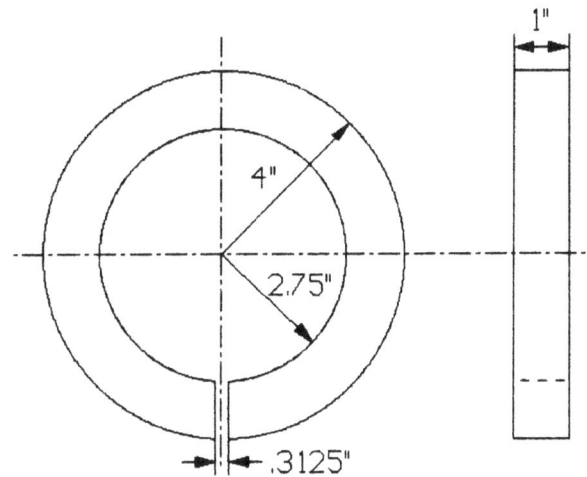

Figure 9-5 Ring dimensions.

2) Cut a slot about .3125" wide in the ring. Be careful to keep cut ends of the ring parallel. I did this by first accurately marking the cuts with a pencil. Then I used a hacksaw to cut part way into the metal on each slot end on each slot end's four sides. After partially cutting the ring this way it was easier to finish the cuts without having the hacksaw drift away from the proper cut. Finally a thin file was used to smooth the ring ends and compensate for any slight hacksaw blade drift.

3) Clean the ring of any loose paint or rust with a wire brush. Then wrap a layer of black plastic electrical insulating tape around the ring and over the slot ring tips. This tape will electrically insulate the coil wire from the ring and will reduce likelihood of accidentally scraping insulation off the magnet wire during coil winding.

4) Carefully and neatly wind 1150 turns of magnet wire around the ring as shown in Figure 9-6. I used about 520 feet of #22 AWG magnet wire. If larger wire such as #20 or #18 AWG is available to you it will work and will result in cooler field coil operation.

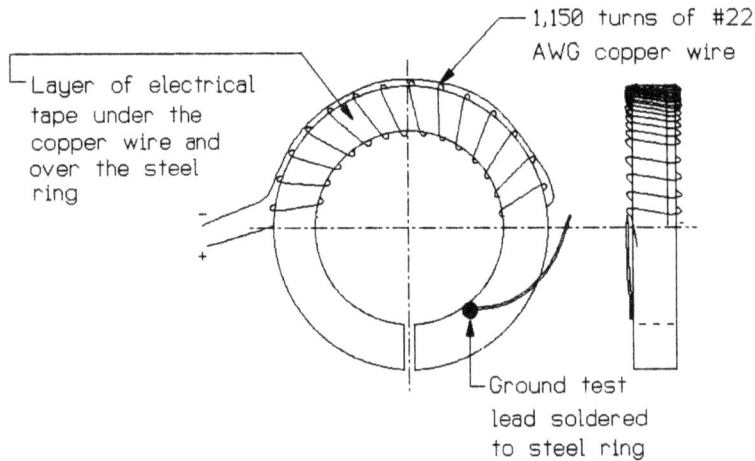

Figure 9-6 Winding the field coil.

5) After the winding is wound wrap a protective layer of black plastic electrical tape over it.

6) Remove the black plastic electrical tape that is over the ring slot tips and over the lower half of the ring. Then, solder an insulated copper lead to a corner of the ring. Use plumbers solder flux when you solder the copper wire to the steel ring. This lead will be used to test if the ring is grounded to the salt water. The ground test lead is shown in Figure 9-6.

7) Paint several layers over the lower half of the ring and ring slot tips. After the paint has dried, tightly wind a layer of black plastic electrical tape over the ring slot tips and the lower ring. The paint and tape will electrically insulate the steel ring from the salt water.

8) Make electrodes from plastic insulated #14 AWG single conductor wire. Strip back each wire's terminal end insulation about 3/8". Strip back the electrode end insulation by the width of the ring. On my ring that was 1". Now use a hammer and small anvil to flatten the copper wire tip into a flat electrode that is a little narrower than the gap. Then bend the wire around so that the flattened wires fit into the gap. Paint the outer side of each electrode. Tape the electrodes to the ring. An electrode is shown in Figure 9-7. The electrodes in place are shown in Figure 9-8.

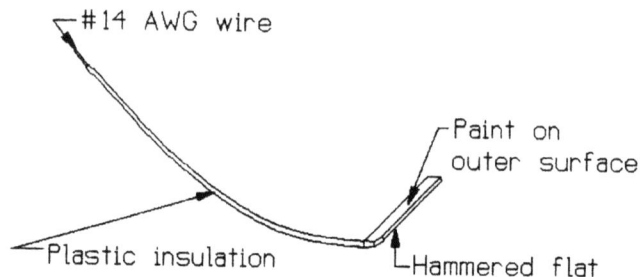

Figure 9-7 One electrode.

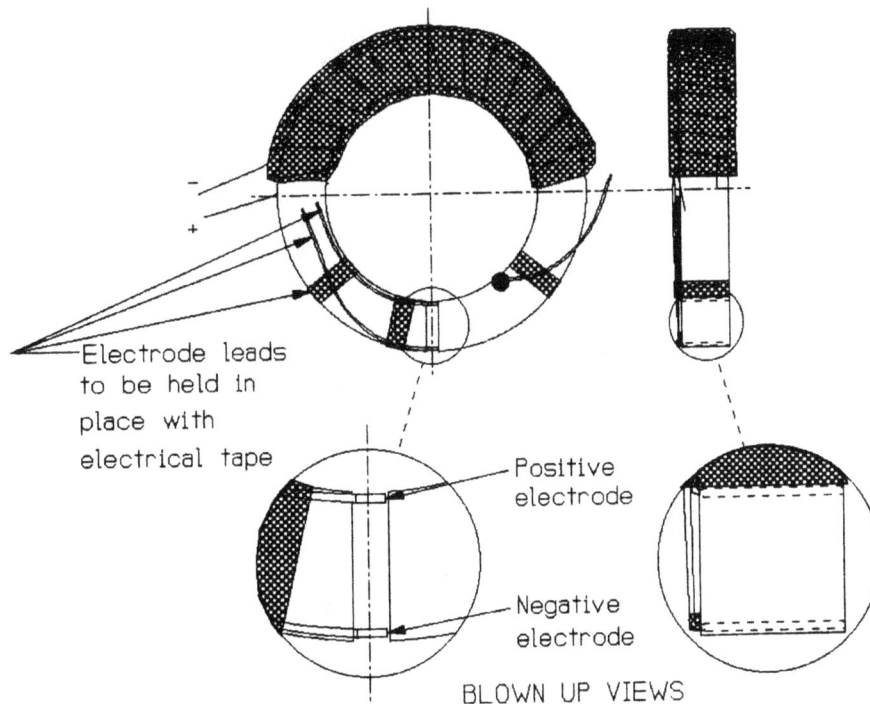

Figure 9-8 Electrodes in place. The black electrical tape that covers the lower ring is not shown here so that the leads are more visible.

9) Test the electromagnetic coil by first measuring its resistance with an ohmmeter. It should be about 8.5 ohms, if #22 AWG wire was used. Measure the resistance between the coil and the ground test lead. There should not be a connection, so the resistance should be very high. If the resistance is not very high there must be a short circuit. Then, the coil must be rewound more carefully. If the resistance readings are correct apply about 12 volts DC to the coil. You can feel the strength of the magnetic field with a screwdriver held over the slot. Note how hot the coil becomes. If the coil becomes too hot it will burn the insulation off the coil wire and short out the coil. As a rule of thumb, when the coil feels moderately warm, the insulation is not being damaged. If the coil feels too hot to touch then shut off power to the coil immediately, the insulation is in danger of being damaged.

10) Put 1 1/2 quarts of fresh water and 7 teaspoons of table salt into the casserole dish. (The ratio of 1 1/2 quarts of fresh water to 7 teaspoons of salt is close to the ratio found in sea water.) Suspend the ring so that its slot and electrodes are totally immersed in the salt water. This is shown in Figure 9-9.

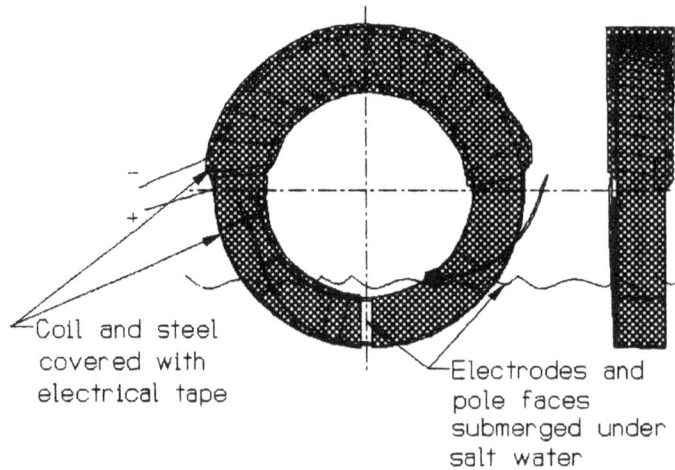

Figure 9-9 Electrodes and gap immersed in salt water.

11) Test the ring for isolation from the salt water. Attach one lead of an ohmmeter to the ground test lead and the other lead of the ohmmeter to a piece of bare metal that is immersed in the salt water. There should not be a connection, so the resistance should be very high. If it is not very high then re-insulate the ring.

12) Connect the DC sources to the coil leads and electrode leads as shown in Figure 9-10. Instrument your circuits with ammeters and voltmeters. If you have variable DC supplies set the supplies to a low voltage and turn them on. Provided that all ammeter and voltmeter readings seem reasonable then increase the voltages of each supply. If you have fixed DC supplies then you will have to turn on your supplies without the safety feature of slowly increasing voltage. When the voltages to the coil and electrodes are about 12 volts DC you should see a stream of bubbles being pushed out of the slot. The bubbles are following the flow of water through the slot. Your MHD system is working.

BE CAREFUL TO NOT GENERATE THE GAS BUBBLES INTO A POORLY VENTILATED AREA! HALF THE BUBBLES ARE CHLORINE, A POISONOUS GAS. HALF THE BUBBLES ARE HYDROGEN, A POTENTIALLY EXPLOSIVE GAS.

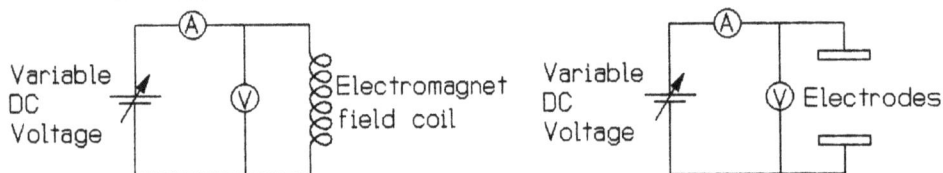

Figure 9-10 Power circuits for the experimental MHD system.

*** DC POWER SUPPLIES

The experimental MHD propulsion system plan calls for two relatively powerful DC power supplies. There may be a problem in finding a source that is powerful, inexpensive, and adjustable. The DC field coil in this project plan will require 1 amp DC at 10 volts DC for long term tests. For short term tests that supply should have a capability of about 6 amps DC at 60 volts DC for tests of a few seconds duration. The DC source for the electrodes should be capable of 1 amp DC at 12 volts DC for long term tests. For short term tests the DC electrode source should be capable of about 4 amp DC at 50 volts DC for tests of a few seconds duration.

Refer to chapter 7 for descriptions of power supplies that could power your experimental MHD system.

*** TESTING YOUR EXPERIMENTAL MHD PROPULSION SYSTEM

Once you have your MHD system operating there are many experiments you can do. For example, you could vary the voltages to the electrodes or coil, vary the salt concentration, find a way of measuring the flow, try different electrodes, etc.

*** REFERENCES

Dane, Abe, "Jet Ships," *Popular Mechanics*, Vol. 167, No. 8 (Aug. 1990), pp. 60-62.

Fine, John C., "Jet Propelled by Magnets:," *Sea Frontiers*, Vol. 37, No. 5 (Oct. 1991), pp. 40-43.

Normile, Dennis, "Superconductivity Goes To Sea," *Popular Science*, Vol. 241, No. 5 (Nov. 1992), pp. 80-85.

10. INDUCTION HEATER

Induction heating is the heating of a conducting and/or magnetic material by changing the amount of magnetic flux through it.

In nonmagnetic materials, induction heating is due only to electric eddy current resistance heating. The electric eddy currents in the material are caused by changing the amount of magnetic flux through the material. This is shown in Figure 10-1.

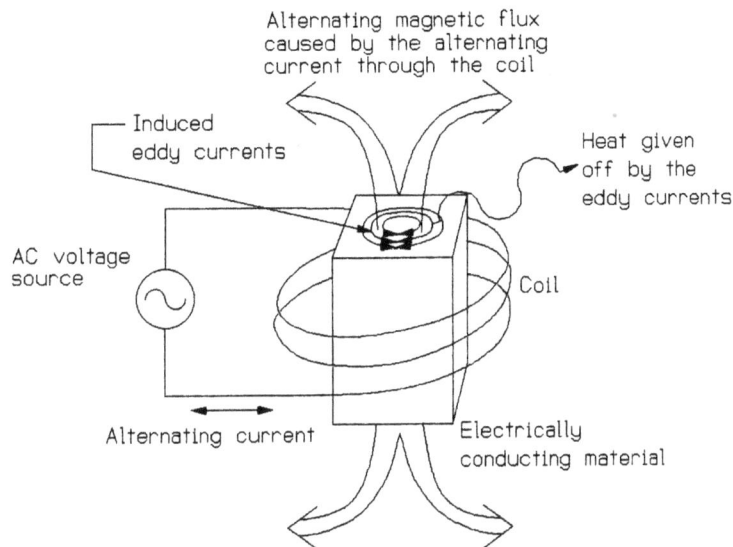

Figure 10-1 Induction heating by electric eddy currents induced in a conducting material. Eddy currents are induced by the alternating magnetic flux of the induction coil. The induction coil produces its flux with the alternating electric current going through it. The frequency of alternation is the same for the induction coil current, magnetic flux, and eddy currents.

In magnetic materials, such as ordinary carbon steel, induction heating is caused by magnetic hysteresis heating as well as electric eddy current resistance heating. Magnetic hysteresis is the lagging of a material's magnetization after a magnetizing force is applied to the material. Every material that can be magnetized has some degree of magnetic hysteresis. In an alternating magnetic field, a magnetic material will be magnetized in one direction, demagnetized, and then magnetized in the opposite direction in a repeating cycle. Materials lose energy as heat every time they are magnetized or demagnetized. The heating that takes place every time a material is magnetized or demagnetized is called magnetic hysteresis heating.

The power of electric eddy current resistance heating caused by induction is proportional to the square of the product of the frequency of the magnetic flux times the amount of magnetic flux. The power of magnetic hysteresis heating caused by induction is proportional to the product of the frequency of the magnetic flux times the amount of magnetic flux.

*** WHERE ARE INDUCTION HEATERS USED?

Induction heaters are used in many different industries. Large induction heaters capable of supplying megawatts of heat are used to heat-treat large billets of steel. Smaller induction heaters with kilowatts of heating capability are used to braze metals together, anneal metals, harden metals, and temper metals. Other induction heaters are used to cure and dry special paints and plastic parts (in a nonconducting material like plastic, a small amount of metal is added so that induction heating can take place). Induction heaters are also used in the processing of semiconductors.

The largest induction heaters, those used to heat steel billets, use low frequencies. Some operate at the power transmission frequency of 60 Hz. Other large induction heaters, using high frequency alternators, produce frequencies up to 10 kHz.

Higher frequency induction heaters produce frequencies up to 10 MHz with SCR (Silicon Controlled Rectifier) or vacuum tube circuits. It is interesting that vacuum tubes are still used in some modern induction heaters, as they are in some modern radio transmitters.

The depth of penetration of the heat produced by an induction heater is inversely proportional to the alternation frequency produced by the induction heater. A low frequency of 60 Hz can be expected to heat a thick billet of steel equally throughout. A high frequency in the MHz region will only heat a material near its surface. One interesting application of this effect is the induction heat treating of engine cylinder head valve seats. By using high frequencies and properly sized induction coils, only the metal surface near the valve seats is heat treated, not the whole cylinder head.

*** MANUFACTURERS OF INDUCTION HEATERS

Following are a few of the manufacturers that offer free informative sales literature on induction heaters.

Inductoheat
32251 N. Avis Dr.
Madison Heights, Michigan 48701
1-800-624-6297
http://www.inductoheat.net/inc/index.cfm

Lepel Corporation
50 Heartland Blvd.
Edgewood, New York 11717
1-800-548-8520
http://www.lepel.com

Radyne
211 West Boden Street
Milwaukee, WI 53207-6277
1-800-236-8360
http://www.radyne.com

Welduction
22750 Heslip Drive
Novi, MI 48375
1-888-INDUCTION
http:/www.welduction.com

*** EXPERIMENTAL INDUCTION HEATER

The induction heater of this chapter is not designed for commercial applications. It is meant for the person who wants to experiment with induction heating. However, there are many characteristics that it has in common with commercially available induction heaters. Among these similarities are the dangers of burns and fire. It will quickly heat a small piece of steel to temperatures that could severely burn a person or set materials afire. The person building this induction heater should remember to be as careful with it as he would with a hot soldering iron.

The experimental induction heater can be explained by breaking its circuit up into stages. The stages are explained as follows and visually shown in Figure 10-2.

1) The first stage is the power supply. In the power supply 120 volts AC at 60 Hz is reduced to two 25.2 volts AC 60 Hz voltages. Each of the 25.2 volts AC 60 Hz voltages is rectified to DC by the bridge rectifiers and then each of the DC voltages is smoothed by the 2,200 µF capacitors.

2) The second stage is the complimentary pair power transistor amplifier and signal generator. The signal generator supplies a high frequency, low voltage, and power input signal to the power transistor amplifier. The power transistor amplifier receives DC power from the power supply. It then amplifies the low power input from the signal generator to a much higher power output.

3) The third stage is the induction coil and series capacitor. They receive high frequency AC voltage from the output of the power transistor amplifier. The frequency of the AC voltage is set so that the series combination of the induction coil and capacitor is in resonance. That way maximum current is passing through them and the maximum voltage is across them. Because of the series resonant condition, the voltage across the induction coil is about six times higher than the voltage supplied by the power transistor amplifier.

4) The fourth stage occurs inside the induction coil. (It is not shown in Figure 10-2.) The high frequency currents passing through the induction coil produce high frequency magnetic fluxes inside the coil. These magnetic fluxes produce eddy current losses and magnetic hysteresis losses in a piece of steel inside the induction coil. The losses cause heating.

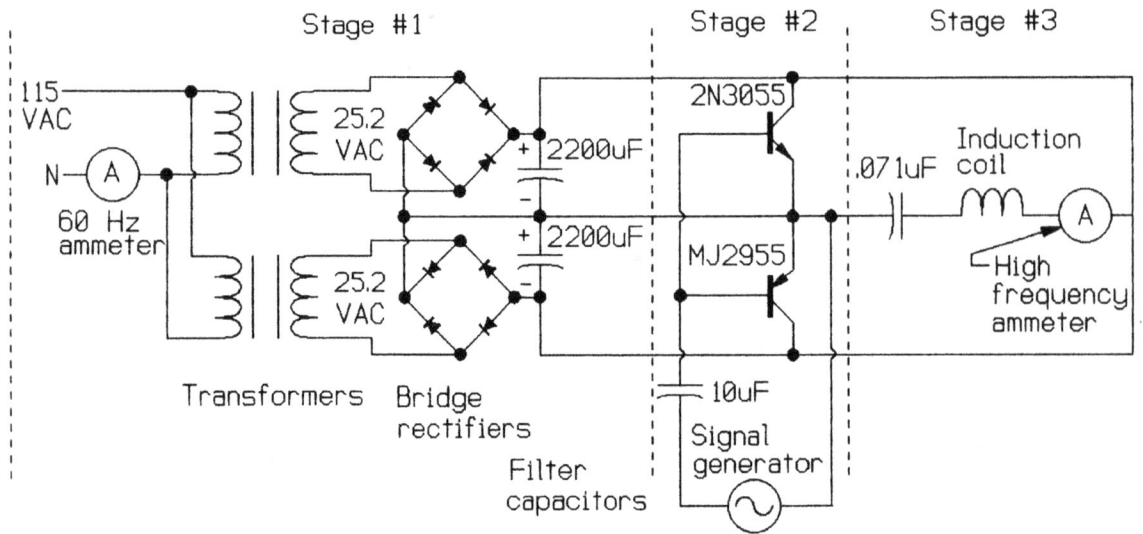

Figure 10-2 Induction heater circuit separated into circuit stages.

*** CONSTRUCTION OF THE EXPERIMENTAL INDUCTION HEATER

BILL OF MATERIALS

ITEM	QUANTITY	WHERE OBTAINED
Pyrex test tube 6" long, 7/8" outside diameter	One	Chemist's supply store
#22 AWG copper magnet wire	40 ft.	Electric motor repair shop or electronics parts store
2N3055 NPN 115 W power transistor	One	Electronics parts store
MJ2955 PNP 150 W power transistor	One	Electronics parts store
Mica electrical insulators for the power transistors	Two	Electronics parts store or scrapped equipment
Heat sink and transistor mounting hardware	One	Scrapped equipment

.071 µF metal film capacitor with at least a 250 volts AC rating	One	Electronics parts store or scrapped equipment
10 µF or greater electrolytic capacitor with at least a 10 volts DC rating	One	Electronics parts store
2200 µF or greater electrolytic capacitors with at least a 35 volts DC rating	Two	Electronics parts store
AC to DC bridge rectifiers, with at least a 50 volt DC and 4 amps DC rating	Two	Electronics parts store
120/25.2 volts AC transformers with a 2 amps rating on the 25.2 volts AC output	Two	Electronics parts store
Assorted colors of #18 or #20 AWG plastic insulated copper wire	Several feet	Electronics parts store
Power cord and outlet plug	About 6 feet long	Electronics parts store or scrap wire
1 3/8" finishing nails for terminals	Six	Hardware store or scrap
Solder and plumber's solder flux	Short piece of solder, dab of solder flux	Hardware store
Black plastic electrical insulating tape	Less than 1 roll	Hardware store
Perforation board for mounting parts, at least 4 ½" x 2 ¼"	One	Electronics parts store
Tall ceramic coffee cup for holding the Pyrex test tube	One	Flea market or store
Wooden mounting board 9 ¼" x 12" x ¾"	One	Lumber yard or scrap
Flat headed wood screws for mounting the transformers and perforation circuit board to mounting board	Five	Hardware store
Large staples for holding the power cord to the mounting board	Two	Hardware store

Epoxy glue for holding the heat sink and coffee cup to the mounting board	One small pack	Hardware store
Alligator clip leads	Four	Electronics parts store
Signal generator capable of 10,000 to 100,000 Hz with a voltage output of 5 volts and a 50 ohm internal impedance	One	Borrow from a school or industrial laboratory or buy from a used equipment flea market. New signal generators are expensive.
0 to 10 amps 60 Hz AC ammeter (a DVM multimeter with scales in this range will do)	One	Electronics parts store
0 to 10 amps AC ammeter capable of reading currents with frequencies of 65,000 Hz (some modern DVM multimeters will do this)	One	Electronics parts store

CONSTRUCTION PROCEDURE:

1) Study Figures 10-2 to 10-4.

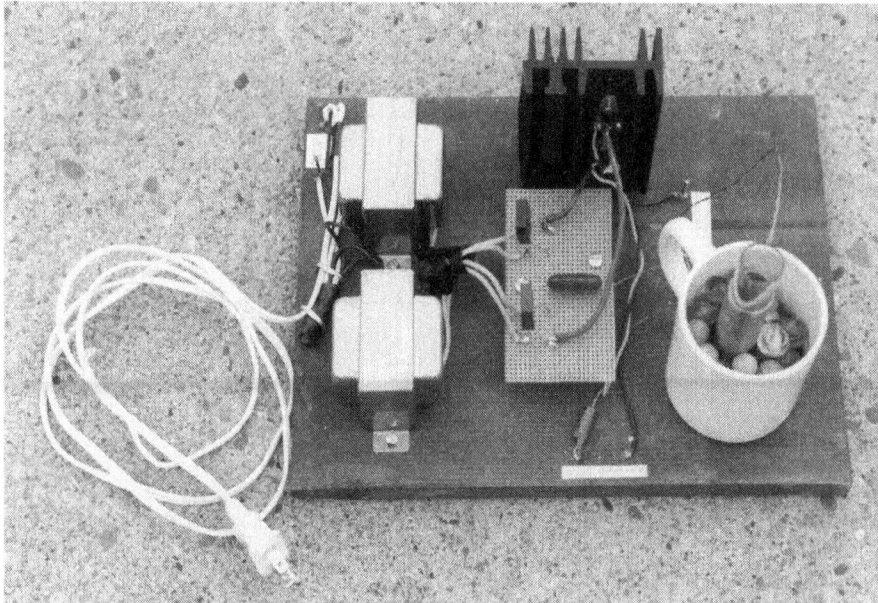

Figure 10-3 Photograph of the induction heater. The signal generator and meters are not shown. Notice that the filter capacitors are under the circuit board. This is to decrease the danger to nearby people in the unlikely event that a capacitor would explode.

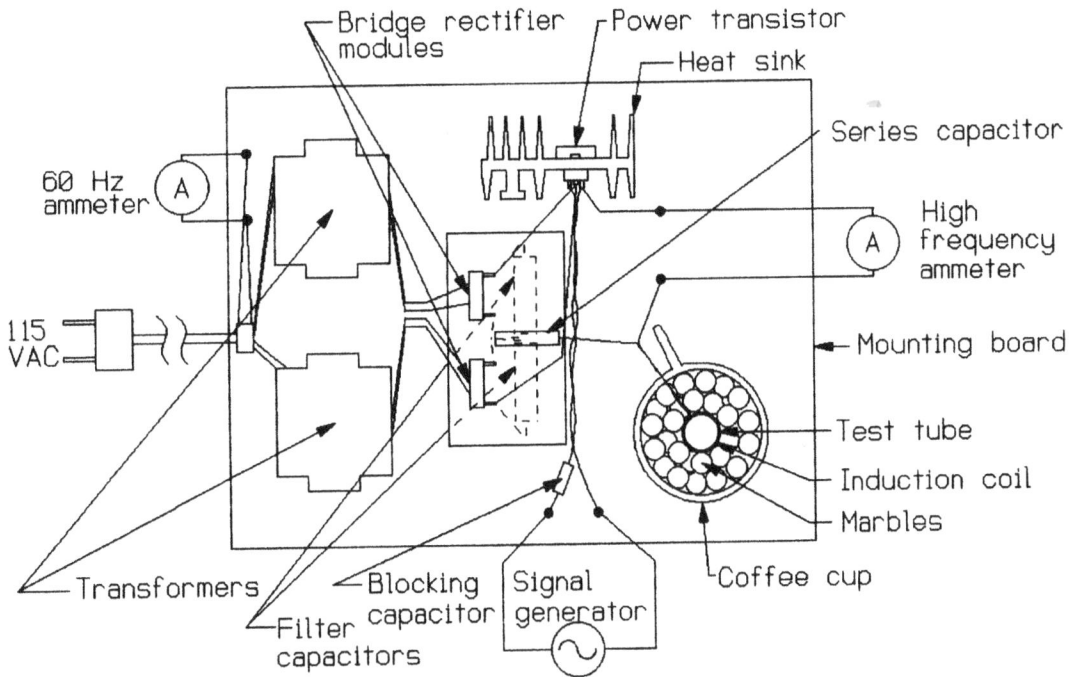

Figure 10-4 Physical drawing of the induction heater.

2) Have the following tools available.

> Electric drill, assorted bits, and a center punch
> Wood saw
> Hacksaw
> File
> Sandpaper
> Hammer
> Screw driver
> Soldering gun
> Pencil
> Carpenter's square

3) Gather the needed parts and materials.

Be certain that you can get a signal generator. As mentioned in the bill of materials, signal generators are expensive. If you do not have one or easy access to one, this project may be too expensive.

4) Constructing the power supply stage.

Mount the transformers on the mounting board with wood screws. Connect the input leads to the transformers in parallel and connect those leads to the power cord. Solder the leads together and cover the connections with electrical tape. Cut one power cord lead near the taped connections. If you have a polarized power plug cut the lead that goes to the neutral. On a two-prong plug, that lead is the one that goes to the larger plug prong. On a three-prong plug that prong is on the right as seen looking from the ground prong. Skin the insulation from the cut lead and wind the skinned ends around clean steel finishing nails that have been driven into the mounting board. Steel finishing nails are used as convenient and inexpensive terminals for connection to alligator clip leads. Coat the nails and the stripped wire with plumber's solder flux. The solder flux will aid in making a good solder joint between the copper wire and the steel nail. Solder the leads in place. Use large staples to hold the power cord to the board. Do not puncture the cord with the staples.

The transformer section should now be tested. Be careful to keep the transformer secondary lead ends from touching each other. Connect an AC ammeter across the finishing nail terminals with alligator clip leads. Plug in the power cord. The ammeter should read about .15 amps. Each transformer should heat up and hum slightly. The output of the transformers' secondaries should be about 28 volts AC (It is higher than 25.2 volts AC because the transformer is not loaded down to its rated current here).

Unplug the power cord.

Following the schematic diagram of Figure 10-5, arrange the two bridge rectifiers, two filter capacitors (2,200 µF) and one series resonant capacitor (.071 µF) on the perforation board. Be careful to connect the polarized electrolytic filter capacitors with their positive terminals connected to the positive terminals on the bridge rectifiers. Mount the filter capacitors under the perforation board to reduce the danger to people in the unlikely event a filter capacitor would explode. See page 60 for information on exploding electrolytic capacitors. When possible, insert the leads of the components through the perforation board. It may be necessary to use a drill to enlarge some of the board holes. Reduce the use of jumper leads as much as possible. Neatly solder the components together. Visually double check that the circuit agrees with the schematic diagram and that there are no accidental dripped solder shorts.

Neatly solder the transformer output leads to the AC inputs of the bridge rectifiers.

Plug in the power cord again. As before, the transformer should be using about .15 amps. Connect a DC voltmeter to the output of the bridge rectifiers. It should measure about 39 volts DC across each output. Connect an AC voltmeter across the outputs of the bridge rectifiers. In this unloaded state, the AC voltage measured should be very small.

5) Constructing the complimentary pair power transistor amplifier stage.

Mount the two power transistors on a single heat sink using mounting hardware and mica electrical insulating spacers. When mounted to a single heat sink, the metal case of the transistors should not make metal to metal electrical contact with the metal heat sink. On my circuit the heat sink, the mounting hardware, and the mica electrical insulating spacers all were salvaged from a scrapped stereo amplifier. After assembly, use an ohmmeter to check that the resistance between the metal can of each transistor and the heat sink is infinite.

If you do not have the materials for a single heat sink, a pair of heat sinks can be used. That way you would use two electrically isolated heat sinks, one heat sink for each power transistor. When using two heat sinks the mica spacers are not necessary, although without them the heat sinks will be electrically "live."

Using #18 or #20 AWG plastic insulated wire, solder the connections shown in the schematic diagram of Figure 10-5 for the amplifier stage. Solder a lead from the emitters of the power transistors to a finishing nail terminal. In all wiring, use the shortest wires possible. Shortness produces neatness and neatness reduces circuit errors. Visually double check that the circuit agrees with the schematic diagram and that there are no accidental dripped solder shorts.

6) Constructing the induction coil and series capacitor stage.

The induction coil is made by winding a single layer of #22 AWG copper magnet wire onto the Pyrex test tube. Pyrex is used instead of ordinary glass because it can better withstand temperature differentials. About 130 turns of wire should be neatly wrapped around the tube. To eliminate the need for taping or gluing down the wire to the tube, wind the wire over itself for one wire end and then twist tie it to the other end of the wire. Twist the lead wires together and be certain to leave plenty of extra lead length to connect to the series capacitor and the finishing nail terminal.

Put the induction heater coil and tube into the tall coffee cup and position it on the mounting board.

The series capacitor was already mounted on the power supply perforation board. Solder the induction coil to the series capacitor and to a finishing nail terminal located by the terminal connected to the power transistors.

Again, visually double check that the circuit agrees with the schematic diagram.

Test your induction heater, following the procedure given in the next section. If it operates properly, then permanently attach the parts to the mounting board. Use epoxy to attach the heat sink(s) and coffee cup and use a wood screw to attach the perforation board.

*** OPERATING YOUR INDUCTION HEATER

Connect the signal generator to the signal input terminals. Connect the 60 Hz 10 amps AC ammeter to the terminals located at the input to the transformers. Connect the high frequency 10 amps AC ammeter to the terminals connected to the induction coil and the amplifier output. Turn on the signal generator with its frequency set to 60,000 Hz and its voltage output set to zero.

Plug in the power cord. Look at the current readings on the two ammeters. The input current should be the same as it was in the power supply tests, about .15 amps. The output current should be close to 0 amps. Slowly increase the voltage of the signal generator. The current readings on each meter should increase. Take care to see that the input current does not increase beyond the rated maximum current allowable by the circuit components. In my circuit the transformers' ratings limited the maximum steady state input current to .84 amps (.42 amps input per transformer).

Increase the signal generator voltage until the transformer input current input is at its maximum steady state value (.84 amps on my circuit) or until it stops increasing. Note the current measured on the high frequency ammeter. It should now have some value in the tenths of amps. Now slowly increase the frequency being produced by the signal generator to greater than 60,000 Hz. Note if the currents increased or decreased. Next decrease the frequency from 60,000 Hz and again note if the currents increased or decreased.

Maximum magnetic flux occurs in the induction coil, when the maximum current is going through it. This will occur at the resonant frequency of the induction coil and series capacitor. Adjust the signal generator's frequency to find that resonant frequency. Adjust the signal generator's output voltage up or down so that the current input to the transformers is close to, but no more than, the maximum steady state input current (.84 amps on my circuit).

With the same signal generator frequency and voltage settings, place a piece of steel in the test tube for about 10 seconds. When the steel is in the test tube, the current readings will drop. The steel changes the inductance of the induction coil and by doing so, changes the magnitude of current passing through it. When the steel is removed, cautiously check to see if it has been heated. If it is hot, your induction heater is working.

The signal generator output voltage and frequency can be tuned to produce a maximum output current and heating when the steel piece is in the induction coil.

*** INDUCTION HEATER DESIGN TIPS AND IMPROVEMENTS

For long term operation, an ordinary household fan blowing air across the heat sink(s) and series capacitor will allow the induction heater to produce more power. If the induction heater is only running for short periods, a fan is not necessary.

The induction coil can be held upright in the tall coffee cup by placing glass playing marbles around it.

The induction heater can put more heat into the material being heated with the use of induction coils with more or less turns and series capacitors with more or less capacitance.

An induction coil with a smaller diameter, but with the same length of coil wire and therefore more turns, can produce a greater magnetic flux density.

In the procedure above, the power transistors are used in their active regions. Transistors operating in the active region produce the smoothest output waveforms, but also produce the greatest heat losses. In the active region, the transistor behaves like a variable resistor. It can become very hot, just like a resistor. If your signal generator can do it, increase its signal to closer to the 4.9 volts rms maximum that the transistors of this circuit can withstand. Using a greater input voltage will cause the power transistors to operate in the active region for only brief moments. Most of the time the transistors will either be off or fully conducting, like a switch. By operating this way, the power transistor amplifier will be cooler and more power will be available for the induction coil.

*** DANGER FROM OPERATING THIS INDUCTION HEATER

The induction heater constructed here can heat pieces of metal to temperatures that can cause fires or severe burns. However, it is not very powerful. It can only produce about 25 watts of heat inside the coil. The heat it can generate in objects outside the coil is slight.

Higher power induction heaters are more dangerous. The University of North London has this to say about induction heaters in their 1996 Health and Safety Bulletin No. 12.

"Any circuit brought near to an induction heater will receive energy and heat up extremely rapidly. No ring or other metal trinket should be worn in the vicinity of an induction heater, nor should any metal be held in the hands as burns may occur in fractions of a second. In general, no part of the body should come with two feet of an induction coil. Personnel with metal bone pins and the like are particularly vulnerable and should avoid such heaters completely."

*** REFERENCES

Zinn, Stanley and Semiatin S. L., *Elements of Induction Heating* (Metals Park, Ohio: ASM International, 1988).

   This book was written under a contract from EPRI (the Electric Power Research Institute). It was designed for industrial readers. It is a book that you might want to look at, but probably would not want to purchase. Library of Congress number: TK4601 .Z56 1988.

Davies, John and Simpson, Peter, *Induction Heating Handbook* (London: McGraw-Hill, 1979).

   This book is also designed for industrial readers. Library of Congress number: TK4601 .D38.

*** INDUCTION HEATER EXPERIMENT

OBJECT

   To determine the power output of the induction heater as it is used to heat a piece of steel.

DISCUSSION

   There are many variables that determine how much heat a particular induction heater can put into a piece of steel. Among these are the type of steel, the shape and weight of the steel, temperature of the steel, and the rate at which heat is lost from steel to the surrounding environment.

   The power determined in this experiment is specific to the particular piece of steel used and the induction heater used.

   The procedures of this experiment are useful when making comparisons between different induction heaters.

MATERIALS

   Induction heater, signal generator, and meters
   Piece of steel
   Pitcher of ice water
   Stop watch
   Metric rule
   Scale

PROCEDURE

1) Determine the volume and mass of the steel piece that is to be placed in the induction heater test tube. Find the mass in grams and the volume in cubic centimeters. The ratio of the mass to volume of the piece should be about 7.87 grams/cm.3.

2) Determine the inside volume of the Pyrex test tube in cubic centimeters. The total volume of the 6" long, 7/8" outside diameter tube I used is 47.5 cm.3.

3) Prepare the pitcher of ice water. The pitcher should contain liquid water and ice cubes. Both the water and ice cubes should be at the freezing temperature of water, $0^{\circ}C$ ($32^{\circ}F$).

4) Put the steel piece in the pitcher to cool it to $0^{\circ}C$.

5) Cool the test tube to near to $0^{\circ}C$ by filling it a couple times with ice water and then pouring the water out.

6) Ready the induction coil for operation.

7) Put the steel piece in the test tube and then pour the liquid ice water (no ice cubes) into the test tube. Now everything in the test tube is close to $0^{\circ}C$.

8) Turn on the induction heater to full power and simultaneously start the stop watch.

9) The magnetic fields of the induction coil will heat the piece of steel and the piece of steel will heat the water surrounding it.

10) After some time the water will boil next to the piece of steel, but the steam bubbles will recondense before reaching the top of the test tube. Write down that time. Soon after the temperature of the whole mass of water will rise to the boiling temperature. Then the bubbles of steam will bubble to the surface. This is the time at which all of the water has reached the boiling temperature. Stop the stop watch and record the time. Turn off the induction heater.

11) The average power received by the steel piece can be calculated with the following equation.

$$Power(watts) = \frac{Water\ Mass(gms.)*419 + Steel\ Mass(gms.)*49}{Time\ to\ Boil(seconds)}$$

The following may be useful in your calculations:

Mass of water in grams = Water volume in cubic centimeters

= Water volume in cubic inches * 16.4

Mass of steel in grams = Steel volume in cubic centimeters * 7.87

= Steel volume in cubic inches * 129.

11. VAN DE GRAAFF GENERATOR PLANS

These are plans for a simple, inexpensive, and relatively safe 100,000 volt Van De Graaff generator suitable for static electricity experiments and demonstrations.

The principles of the Van De Graaff generator were discussed on pages 21 and 22. The only difference between the generator discussed there and here is that this one lacks a high voltage supply to spray charge onto the belt. This generator uses triboelectrification (electrification by friction) to cause negative charge to travel from the moving plastic belt to the lower electrode. Negative charge is actually rubbed from the plastic belt and deposited on the leather electrode. The end result of this is that the dome becomes positively charged.

*** BILL OF MATERIALS

ITEM	QUANTITY	WHERE OBTAINED
12" frying pan, aluminum is preferable for ease of drilling	One	Flea market
3 quart cooking pot	One	Flea market
2 3/8" outside diameter plastic pipe	17"	Hardware store
1" outside diameter wooden dowel	3"	Hardware store
Fan motor from a large kitchen microwave cooker or phonograph turntable motor, each with motor mounting hardware	One	Scrapped microwave cooker or phonograph
20d nail or a 1/8" outside diameter, 4" long metal rod	One	Hardware store
Suede leather	About 5" x 4"	Palm of an old leather work glove
Thin plywood or smooth paneling	One square foot	Scrap lumber
Aluminum 1" x 1" angle	6"	Hardware store
Aluminum foil	3 square feet	Grocery store
Small pieces of wood about 1" x ½" x ½"	Three pieces	Scrap lumber

#14 AWG bare copper wire, form stripped Romex wire	1 foot	Hardware store
Flexible lamp cord	6"	Hardware store
#18 or #20 AWG plastic insulated copper wire	Several feet	Electronics parts store or scrap wire
Power cord and outlet plug	About 6 feet long	Electronics parts store or scrap wire
Small alligator clip	One	Electronics parts store
Large rubber band or strip of rubber cut from a bicycle inner tube	One	Stationery store or bicycle shop
Electrical solder	Several inches	Electronics parts store
6-32 3/8" long screws with nuts and lock washers	Seven	Hardware store
Plastic electrical tape ¾" wide	One roll	Electronics parts store
Epoxy glue	One small pack	Hardware store

*** CONSTRUCTION PROCEDURE

1) Study Figures 11-1 and 11-2.

Figure 11-1 Photograph of the Van De Graff generator. Note the triboelectric electrode and its grounding alligator clip.

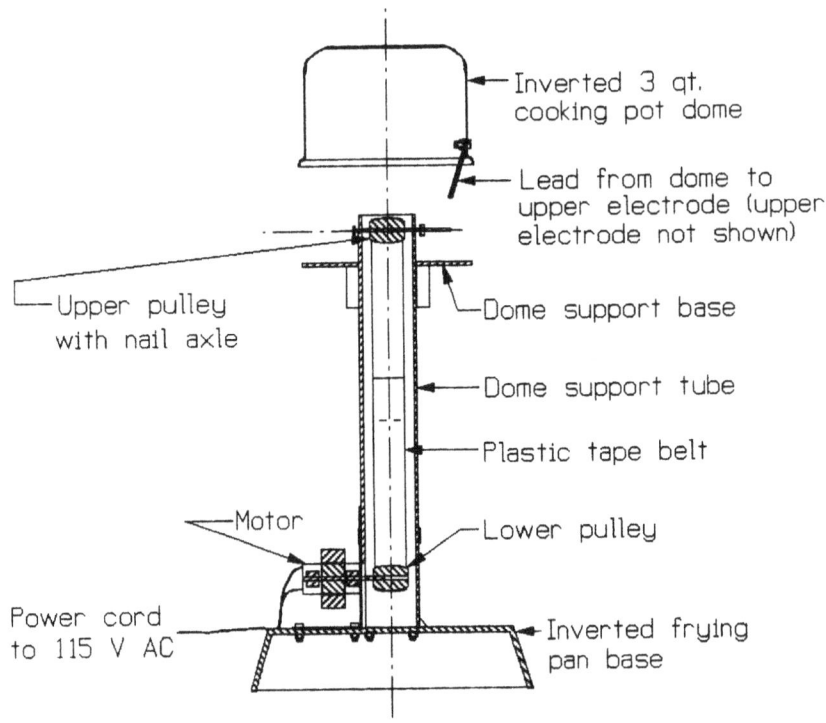

Figure 11-2 Cross-section of the Van De Graaff generator.

2) Have the following tools available.

      Electric drill, assorted bits, and a center punch
      Hacksaw
      File
      Coping saw or electric jig saw
      Screw driver
      Adjustable wrench
      Soldering gun
      File
      Scissors
      Carpenter's square

3) Assemble parts.

    a) Mount the motor on the inverted frying pan as shown in Figure 11-3. Put the shaft of the motor above the middle of the frying pan bottom and parallel to the plane of the frying pan bottom. I used a microwave cooker fan motor with its microwave mounting hardware. Holes were drilled through the motor mounting hardware and frying pan so that the motor could be held down by 6-32 screws, lock washers, and nuts.

Figure 11-3 Photograph of the motor mounted on the frying pan. Note the lower pulley, elastic band, aluminum angles, motor mounting hardware, and leather/aluminum foil electrode.

b) The pulleys are made from 1" diameter wooden dowels. First, cut off several 1 1/2" long lengths of the dowel. Be careful to keep the cuts perpendicular to the axis of the dowel. Make corrections by sanding. Next, carefully mark the centers of the dowels. Then use an electric drill to drill in from each end of each dowel about one half a dowel length. Use a drill bit that is slightly larger that the shaft of the motor. I used a 1/8" drill bit. After drilling half way from each end, run the drill bit the whole way through. If you were skillful and lucky, the drilled hole will be close to the centerline of the dowel.

Select the best two dowels and slide one of them onto the motor shaft. It should fit snugly. Run the motor and use the hacksaw blade as a cutting tool to round and shape a crown onto the dowel. The rounding process will correct for a slightly off center hole. The diameter of the shaped dowel should be about 1/8" less at the ends than in the middle. The crown shape will keep the belt centered. Again running the motor, use sandpaper to smooth the pulley after it is shaped.

Slide off the shaped dowel and repeat the procedure with a second dowel.

The lower pulley is shown in place in Figure 11-3.

c) The dome support tube is made from a piece of plastic pipe that has a 2 3/8" outside diameter and a 17" length. After the pipe has been cut, use the carpenter's square to verify that its ends are square. Make corrections with sandpaper.

At the bottom end of the support tube a slot should be made. The slot should be large enough to allow the tube to be easily positioned over the bottom pulley and the lower electrode. Also, cut a 1" hole 1" into one side of the dome support tube. The hole will be the access for the lower electrode, the triboelectric electrode. The dome support tube is shown in Figure 11-4.

Figure 11-4 Photograph of the dome support tube and dome. Note the slot in the tube, lower electrode hole, aluminum foil covered platform, upper electrode, upper pulley, nail used as an axle, and wire connecting the upper electrode to the dome.

d) To hold the dome support tube in position, aluminum angles are attached to the frying pan and used as positioning rails. A large rubber band is used to hold the dome support tube to the motor mount. The advantage of using a rubber band is the ease with which the tube can be removed from the inverted frying pan base. That is very useful as adjustments are made during construction and when the Van De Graaff is disassembled for moving. The rails and rubber band can be seen in Figure 11-1.

e) The upper pulley is mounted on a shaft made from a 20d nail. The nail goes through the hole drilled through the center of the dowel and through holes drilled through the plastic pipe. To keep the belt centered during operation, the holes drilled through the plastic pipe are each at the same height above the bottom of the dome support tube and the tube ends are perpendicular to the tube axis. A small square of rubber (from an inner tube or large elastic band) is pierced by the nail to hold the nail in the tube. The upper pulley can be seen in Figure 11-4.

f) The Van De Graaff generator belt is made from lengths of black plastic electrical insulating tape. To determine the belt length cut a 1 1/2" wide strip of newspaper. Put the strip over the Van De Graaff pulleys where the belt will be. Then carefully mark the strip so that it could be cut and glued to form a belt. The belt of black plastic tape will be made to the same length as the marked newspaper strip.

Clear off an area along a straight edge of a flat and clean table. Use the newspaper strip to measure off the length needed. Along the table edge, unwind a length of black plastic tape that is five inches longer than the marked newspaper strip. When unwinding the tape be careful to not stretch it and be careful to keep it straight. Use small pieces of tape to secure the length of tape to the table at each end. Unwind a second length of tape sticky side up so that it is stuck to the first length with an overlap of about 1/8". Again use small pieces of tape to secure its ends to the table. Unwind a third length of tape and carefully place it sticky side down onto the sticky side of the first tape. Cut this tape so that is shorter than the lengths of the first and second tapes, but slightly longer than the length indicated by the newspaper strip. Unwind a fourth length of tape so that it covers the second tape and starts and ends at the same locations as the third tape.

Using the paper strip to determine length, squarely trim off the beginning ends of the four tapes. Next squarely trim the other ends of the tapes one and two, the longer tapes. Being careful to keep the tapes straight, stick the tapes one and two back onto themselves. Finally, stick small pieces of tape across each of the joint gaps. The belt is now formed.

g) To install the belt it is necessary to disconnect and lower the upper pulley. I used a bent paper clip to temporarily lower the upper pulley when putting the belt in place.

h) With the belt and pulleys in place, run the motor. Although the belt may slip a little bit at first, the belt should soon be driven by the motor. If the belt slips too much, a shorter belt should be made. If the motor will not turn at all a longer belt should be made. It is possible to adjust the belt length slightly by unsticking it at the joint and removing material or making a larger gap under the small pieces of tape across the joint gaps.

i) A platform holds the upper electrode and the dome. The platform is made of a donut shaped piece of thin plywood with a hole that just fits over the support tube. The outer circumference of the plywood just fits inside the outer lip of the cooking pot being used as the dome. The platform is supported on the tube by three small pieces of wood that have been taped to the tube. The height of the platform on the tube is set so that the top of the upper pulley is halfway between the top of the dome (bottom of the cooking pot) and the platform. The platform can be seen in Figure 11-4.

j) The upper electrode is made from bare #14 AWG copper wire, small gauge copper wire taken from a lamp cord and solder. Twist or tie the small gauge copper wire onto the #14 AWG wire so that the tips of the small gauge wire make a brush. Then solder the small gauge wire in place. Mount the upper electrode to the platform using epoxy glue. The upper electrode should be mounted so the brush tips will touch the belt. This can be seen in Figure 11-4. Solder a length of #18 AWG to the end of the #14 AWG wire so that a solid electrical connection can be made to the dome. At the dome use a machine screw, nut, and lock washer to connect the lamp cord. I connected the wire at one of the holes that was left when I removed the cooking pot's handle.

k) Wrap the platform completely with aluminum foil.

l) Put the dome in place.

m) Solder an alligator clip to a 6" length of #18 AWG wire and attach the other end to one of the screws holding the aluminum angle on the 1" square hole side of the support tube.

n) Lay a piece of aluminum foil onto the suede leather piece. Fold the leather so that it is like a jelly roll with the aluminum foil in the middle. Clip the alligator clip to the aluminum foil. Start the motor. Now gently push the suede leather/aluminum foil combination into the 1" square hole in the side of the support tube. The leather should rub against the moving belt, but should not stop it.

o) Your Van De Graaff is now in operation.

*** OPERATING YOUR VAN DE GRAAFF GENERATOR

Plug in the power cord and the Van De Graaff generator will produce a high static voltage, provided that the air is dry. It can produce arcs up to 1" long to your finger tips and up to 2" long to more rounded objects like a metal pot. In humid air it won't operate. On the days that you find yourself being shocked after walking across a carpet, the air is dry enough.

*** BUILDING HIGHER VOLTAGE VAN DE GRAAFF GENERATORS

The output voltage of a Van De Graaff generator relative to earth potential is proportional to the amount of charge stored on its dome. Its voltage will rise until charge leaks off the dome as fast as the belt supplies more charge. There are many possible design improvements that will raise the dome potential.

First, the Van De Graaff could be operated in drier air. In Pennsylvania, my Van De Graaff operates properly only in the dry indoors winter air.

Second, the dome shape can be made more rounded. Charge tends to leak off a charged object at sharp corners either by arc discharge or by corona discharge. The best dome has no sharp corners and is nearly a perfect sphere. Also, within the limitations of the height of the support tube, a larger dome is better. A larger dome can contain more charge and is less sharp.

Third, the electrodes and belt can be improved. The belt and electrodes could be widened to carry more charge. The speed of the belt could be increased to carry more charge. The belt material could be improved. Another belt material may be able to carry more charge.

Fourth, a high voltage supply can be incorporated into the design. The output voltage and current capacity of the Van De Graaff would be increased by connecting a high voltage DC supply between the ground and a brush type lower electrode. This is shown in Figure 2-10 on page 21. (An added benefit of using a charging supply with a brush type electrode rather than a triboelectric electrode is that the polarity of the dome voltage can be switched by switching the polarity of the charging supply.)

*** WHAT IF YOU NEED A VAN DE GRAAFF GENERATOR, BUT HAVE DECIDED AGAINST MAKING YOUR OWN?

Edmund's Scientific sells reasonably priced Van De Graaff generators. Contact them at:
Edmund Scientific Company
101 East Gloucester Pike
Barrington, NJ 08007-1380

*** DANGER FROM OPERATING THIS VAN DE GRAAFF GENERATOR

With the design given here the shocks one might receive are no more dangerous than the shocks one receives on dry days by walking across a carpeted floor and touching a metal door knob. However, be careful if you use your Van De Graaff to charge a high voltage capacitor, such as a Lyden jar. A high capacity high voltage capacitor could be charged to lethal levels with this Van De Graaff generator.

*** REFERENCE

McComb, Gordon, *Gadgeteer's Goldmine* (Blue Ridge Summit, PA: TAB Books, 1990). Purchase from TAB Books, Division of McGraw-Hill, Inc., Blue Ridge Summit, PA 17294-0850.

This book has plans for a small Van De Graaff generator that uses a high voltage supply to spray charge onto the belt. It also has plans and descriptions of many other interesting electrical and electronic devices. Library of Congress number: TK9965 .M35 1990.

*** VAN DE GRAAFF GENERATOR DEMONSTRATION EXPERIMENT

OBJECT

To use the Van De Graaff generator to demonstrate that your body can carry charge and that like charged objects repel each other.

MATERIALS

Van De Graaff generator
Three strong and inexpensive ceramic coffee cups
One board about 1' x 1' and 3/4" thick.

PROCEDURE

1) Wash and dry your hair so that it is not sticky or greasy.

2) Set the Van De Graaff generator on a table.

3) Arrange the three coffee cups on the floor as electrically insulating supports for the board.

4) Turn on the Van De Graaff generator.

5) Carefully stand on the board and put one of your hands on the dome of the operating Van De Graaff generator.

6) Your hair should stand straight out.

The charge from the dome conducts to you and spreads out across your body and onto each hair. Your body and each hair becomes charged with the same polarity charges. Because like charges repel, each hair repels itself away from the body and away from the other like charged hairs. The result is that your hair stands straight out.

12. ELECTROPLATING WITH COPPER

Electroplating is the deposition of a thin layer of metal onto a conducting surface (usually another metal) by electrolysis. It is used to protect and decorate the underlaying conducting surface. The most common electroplating metals are cadmium, chromium, copper, gold, nickel, silver, and tin. Layers are usually in the thousandths of an inch. Typical products of electroplating are silver-plated tableware and chromium-plated automobile bumpers. Figure 12-1 shows a steel bolt being electroplated with copper.

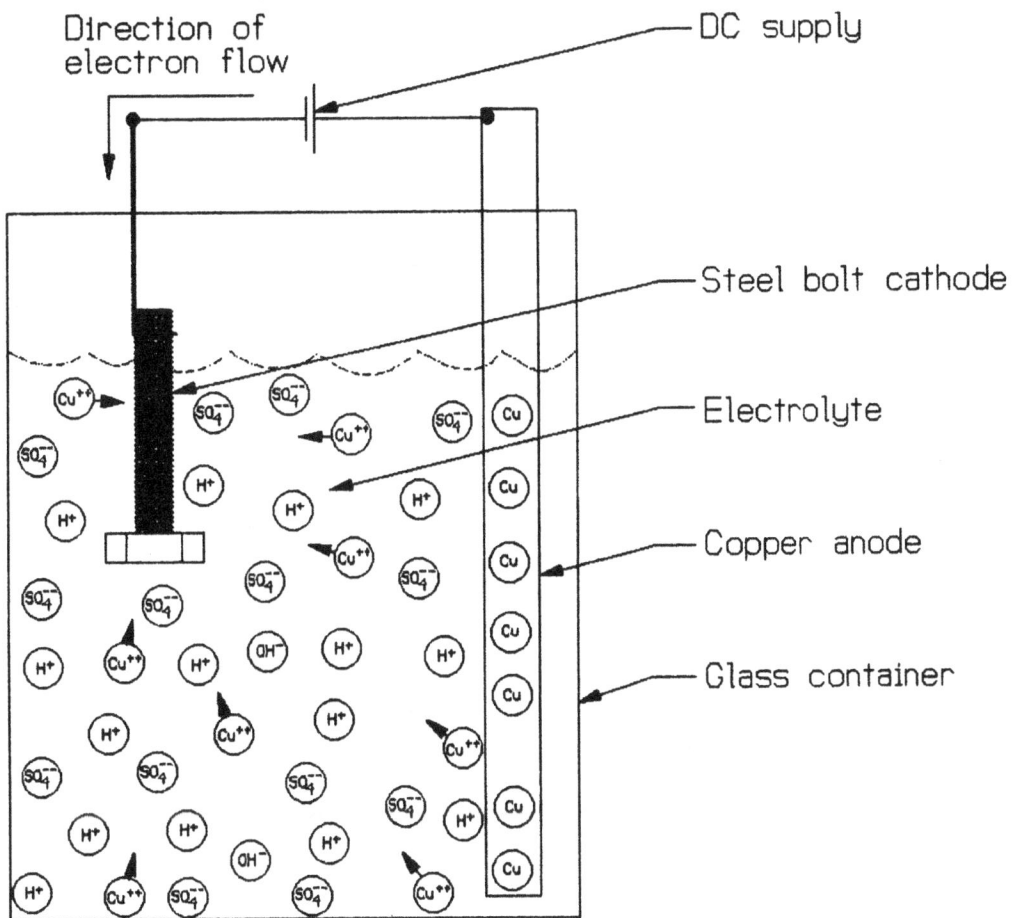

Figure 12-1 Steel bolt being electroplated with copper. The liquid electrolyte solution is composed of copper sulfate, sulfuric acid and distilled water.

Electroplating was invented by Luigi Brugnatelli, an Italian, in 1805. Like all electroplaters until near the end of the 19th century, Brugnatelli used a "cell apparatus". This "cell apparatus" was equivalent to a short-circuited Daniell cell. Electroplating is one of the oldest practical uses of electrical power and is still in common use today.

*** HOW DOES ELECTROPLATING WORK?

The anode is the source of the electroplating metal. It is connected to a positive DC potential. The cathode is the conducting surface that is being plated. It is connected to a negative DC potential. In Figure 12-1, the anode is the copper sheet, the steel bolt is the cathode.

The liquid that carries the metal from the anode to the cathode is called the electrolyte. It is usually made of water and a dissolved metal salt(s). Sometimes, there is also an acid or a base in the electrolyte. In Figure 12-1, the electrolyte is made of water, dissolved copper sulfate ($CuSO_4$), and sulfuric acid (H_2SO_4). In the water solution, the copper sulfate and sulfuric acid separate into ions. The copper sulfate becomes positive copper ions and negative sulfate ions ($CuSO_4 => Cu^{++} + SO_4^-$). The sulfuric acid and water become positive hydrogen ions and negative sulfate ions ($H_2SO_4 => 2H^+ + SO_4^-$). Some of the water also dissociates (separates) into positive hydrogen ions and negative hydroxide ions ($H_2O => H^+ + OH^-$).

When electroplating is taking place, the cathode supplies electrons to the positive metal ions that touch it. These metal ions then become neutral metal atoms that remain on the cathode. At the surface of the anode, neutral metal atoms are stripped of their electrons, which converts them into positive metal ions. These metal ions then float off into the electrolyte. In Figure 12-1, the cathode supplies electrons to the positive copper ions and positive hydrogen ions that touch it. The copper ions become neutral copper atoms and are deposited on the cathode. The hydrogen ions become neutral hydrogen molecules and form into hydrogen gas bubbles. Meanwhile, at the surface of the anode, neutral copper atoms give up electrons and combine with the electrolyte's sulfate ions, thus resupplying the solution with copper sulfate. At the same time, electrons are received at the surface of the anode from hydroxide ions that touch the anode. The hydroxide ions then become neutral oxygen and water molecules. The oxygen molecules form into oxygen gas bubbles.

During electroplating, the higher the concentration of metal ions is in the electrolyte, the less will be the production of hydrogen and oxygen gas. In a well run electroplating system, no visible gas bubbles are produced.

*** CONSTRUCTION OF A CuSO₄/H₂SO₄ COPPER ELECTROPLATING SYSTEM

BILL OF MATERIALS

ITEM	QUANTITY	WHERE OBTAINED
One gallon glass pickle jar	One	Restaurant
Low voltage variable DC power supply, similar to those described in chapter 7. The supply capability should be 0 to 4 volts and up to 1 amp.	One	See chapter 7
Copper sulfate	One pound	Hardware store, copper sulfate is often sold for use as a root killer for clearing sewer pipes
Dilute sulfuric acid, 1.25 specific gravity or Concentrated sulfuric acid, 1.84 specific gravity	200 milliliters or 60 milliliters	Automobile parts store or Chemical supply company
Distilled water or rain water	3 quarts	Grocery store or sky
Sheet of copper, copper pipe, or bare copper wire	About ¼ pound	Scrap material
Bucket of water for emergency rinsing	2 gallons	Hardware store and kitchen faucet
Alligator clip leads	Three	Electronics parts store
0 to 10 amp DC ammeter (a multimeter with scales in this range will do)	One	Electronics parts store

SAFETY WITH CHEMICALS

Skin contact with some chemicals will cause chemical burns. Many chemicals are poisonous. Combining the wrong chemicals can cause poisonous fumes or explosions.

Copper sulfate may cause severe skin and eye irritations. It is harmful if absorbed through the skin or inhaled. It is poisonous. It is toxic to fish. If it comes in contact with the eyes or skin, flush with great quantities of water and then call a physician. If swallowed, promptly drink a large quantity of milk, egg white, gelatin solution, or if none of these is available, large quantities of water. Then call a physician. AVOID CONTACT WITH COPPER SULFATE. USE RUBBER GLOVES, A FACE SHIELD, AND A RUBBER APRON.

Sulfuric acid is a dangerous chemical. Contact with the skin will cause chemical burns (the dissolving of flesh). It will produce poisonous vapors and is extremely poisonous to drink. The severity of the damage it produces depends on its concentration. Pure (concentrated) sulfuric acid is more dangerous than the dilute sulfuric acid that is used in automobile batteries. If it comes in contact with the skin, flush with great quantities of water and then call a physician. If sulfuric acid comes in contact with the eyes, flush with great quantities of water for 15 minutes and then call a physician. If swallowed, promptly drink large quantities of water or milk. Follow with milk of magnesia, beaten eggs, or vegetable oil. Then call a physician. AVOID CONTACT WITH SULFURIC ACID AND AVOID BREATHING ITS VAPORS. USE RUBBER GLOVES, A FACE SHIELD, A RUBBER APRON, AND WORK IN A WELL VENTILATED AREA.

Sulfuric acid, when combined with water or bases, will produce heat. In some cases this heat is enough to boil water, causing an acid splattering steam explosion. Bases are a sort of chemical opposite to acids. Lye mixed with water forms a strong base. The combination of lye and sulfuric acid instantly creates a great deal of heat in an explosive and dangerous reaction. Water added to concentrated sulfuric acid will be heated as it combines with the acid. This heat may be sufficient to cause the water to boil and cause the acid to be splattered around by the boiling water. However, it is acceptable to slowly pour acid into water, pouring that way the heat created is dissipated in the water and the water does not boil. DO NOT COMBINE ACIDS WITH BASES. DO NOT POUR WATER INTO ACID, POUR ACID INTO WATER.

Mark containers containing poisonous chemicals as POISON. Neutralize excess sulfuric acid or sulfuric acid spills with baking soda.

KEEP ALL CHEMICALS OUT OF REACH OF CHILDREN.

CONSTRUCTION PROCEDURE:

1) Study Figures 12-2 and 12-3.

Figure 12-2 Photograph of a CuSO$_4$/H$_2$SO$_4$ copper electroplating system.

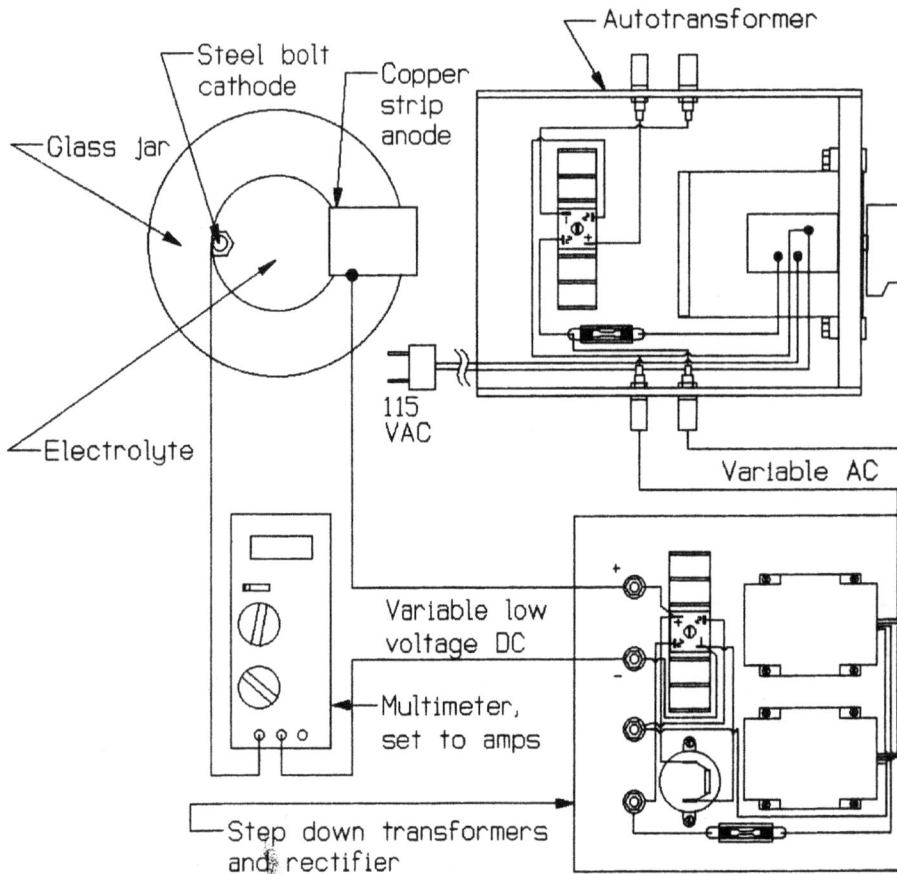

Figure 12-3 Physical drawing of the CuSO$_4$/H$_2$SO$_4$ copper electroplating system.

2) Making the electrolyte.

   a) Fill the 2 gallon bucket with tap water and put it in an easily accessible location. It is to be used for emergency rinsing in case of an acid spill or an accidental contact with acid or copper sulfate.

   b) Put 1 pound of copper sulfate into the gallon jar. Pour 3 quarts of distilled water into the jar. Slowly pour the sulfuric acid into the mixture, either 200 milliliters of automobile battery dilute sulfuric acid or 60 milliliters of concentrated sulfuric acid. Put the water into the mixture before adding acid! DO NOT POUR WATER INTO ACID, POUR ACID INTO WATER. Stir the mixture with a glass rod or a small piece of clean copper pipe.

3) Making the anode. Put the copper anode into the gallon jar so that part of it overhangs the top and the rest is submerged in the electrolyte. As much as possible, shape the copper anode

so that is covers a wide area of the inside surface of the gallon jar. The larger and more symmetrically placed the anode is around the inside of the jar, the more evenly objects will be electroplated.

4) Preparing a metal object for electroplating. It is important for the object to be clean and conductive. An unclean object will not be uniformly plated. Nonconductive spots on the object will not plate. Clean the object thoroughly. Use a fine steel wool, if it won't damage the object. Do a final cleaning with alcohol to clean off any oils.

5) Making the DC supply circuitry. The circuitry described in Chapter 7 was used to create a variable low voltage and current supply. It is shown in Figures 12-2 and 12-3.

*** ELECTROPLATING A SMALL BOLT WITH YOUR CuSO$_4$/H$_2$SO$_4$ SYSTEM

1) Select a small steel bolt. Estimate its surface area in square inches. The maximum allowable electroplating current in amps equals the product of the area in square inches times .035. Electroplating will occur with greater currents, but greater currents often cause excessive gas bubbling, resulting a poor electroplate. Good electroplating can be done with lesser currents. However, the rate of electroplating decreases as current decreases.

2) Clean the bolt. The cleaner and smoother the surface, the better the electroplate will be. Fine steel wool should be used to take off the outer layer of dirt and rust. Then use alcohol to remove any oils remaining on the bolt's surface.

3) Attach the bolt to the negative potential. If you use an alligator clip to connect to the bolt and dip it with the bolt into the electrolyte, the alligator clip will also be plated. Also, at the point of contact where the alligator clip touches the bolt, the bolt will not be electroplated. To save the alligator clip and electroplate the bolt over the bolt's entire surface, hang the bolt so only half the bolt is immersed in the electrolyte. Later hang the bolt from the other end.

4) The longer the bolt remains in the electrolyte and the greater the electroplating current the thicker the electroplate will be.

5) After plating is finished, rinse the bolt off in fresh water.

*** CuSO₄/H₂SO₄ COPPER IMMERSION PLATING

Many metals, such as steel, can be given a thin electroplate by simply dipping them into the copper sulfate/sulfuric acid electrolyte without applying an external voltage. This is called immersion or electroless plating.

In immersion plating, the metal on the outer layer of the object being plated trades places with the copper ions in the electrolyte. Steel would loose a very thin layer of iron atoms to the electrolyte. The iron atoms would be ionized and float out into the electrolyte. At the same time, the copper ions of the electrolyte would take the electrons left by the iron atoms and become neutral atoms on the surface of the steel object.

Immersion plating works with many metal combinations, but only in one direction. Its operation depends on one metal having a greater attraction for electrons than the other. Immersion plating of steel (iron) in a copper sulfate/sulfuric acid electrolyte works. Immersion plating of copper in an iron sulfate/sulfuric acid electrolyte would not work, the copper would not be replaced by iron.

Usually, the quality of the plate produced by immersion plating is poor in comparison with that produced by electrically powered electroplating.

*** ELECTROFORMING AND ELECTROREFINING

Electroforming (also called electrotyping) uses the same electrolysis process as electroplating. However, with electroforming a thick layer of metal is deposited in a mold. The first records of it being done date back to 1838. It is used to make small metal objects.

Electrorefining also uses the same process as electroplating to purify metals. The anode is typically made of a mixture of the desired metal and other metals or materials. The cathode receives only the desired metal. The impurities either go into the electrolyte solution or fall to the bottom of the electrorefining tank.

*** ELECTROPLATING NONCONDUCTING SURFACES

To electroplate a nonconducting surface, it must first be coated with a conductive material.

For the small scale electroplater, the best way of making a surface conductive is with a conductive silver paint. The book by Ammen (see REFERENCES) mentions good electroplating results with DuPont silver conductor paint #4817. However, this silver conductor paint is expensive, it costs about $1.00/gram for small quantities. Contact DuPont Microcircuit Materials at 1-800-284-3382 to find a local dealer.

Another way is to coat the surface with bronzing powder. Bronzing powder is made of fine copper filings. It can be purchased or made. To make it, simply file a piece of copper with a fine file. The surface of the object to be electroplated should be first coated with a thinned lacquer or polyurethane. Then, before the lacquer or polyurethane is quite dry, the bronzing powder should be sprinkled evenly onto it. Enough bronzing power should be sprinkled on so that the nonconducting surface is no longer visible. This method is much less expensive than using conductive silver paint, but the resulting electroplate is coarser.

*** ELECTROPLATING WITH OTHER METALS

Many metals can be used in electroplating. However, the problem with electroplating with most metals is that very toxic chemicals are required. Most require the use of cyanide salts. These cyanide salts are safe only when properly used. With improper handling they are deadly. A cyanide salt in combination with an acid will suddenly produce hydrogen cyanide (hydrocyanic gas), in the same way it is produced in the gas execution chambers.

If the reader is interested in further pursuing electroplating with other metals, he should take some basic chemistry classes and read more about electroplating. Some references are given at the end of this chapter.

*** WHO SHOULD DO $CuSO_4$/H_2SO_4 ELECTROPLATING?

The person doing this should be able to handle sulfuric acid and copper sulfate safely. He should not be clumsy or prone to dropping things. He should at least have a high school chemistry level understanding of the chemical processes involved.

*** WHAT IF YOU NEED TO ELECTROPLATE SOMETHING, BUT DON'T WANT TO DO IT YOURSELF?

There are many small electroplating businesses. Often these businesses specialize in chrome plating for car parts. Call around your area to find out who can help you in your electroplating needs.

*** CuSO$_4$/H$_2$SO$_4$ ELECTROPLATING TIPS AND IMPROVEMENTS

Keep the electrolyte clean, otherwise impurities may appear on the surfaces being electroplated. Professionals use filter systems to do this.

Move the object in the electrolyte during electroplating to obtain a more even electroplate at higher plating currents.

Often professionals use compressed air agitation or pumps to keep the electrolyte evenly concentrated.

Do not operate your electroplating system in a room with poor ventilation. The vapors it gives off are unhealthy and will accelerate the corrosion of all metal objects in the room.

*** REFERENCES

The Electroplater's Handbook, by C. W. Ammen, TAB Books Inc., Blue Ridge Summit, PA copyright 1986.
    This excellent book describes electroplating with many different metals and chemicals. It was written for the beginner who wishes to do small scale electroplating. Library of Congress number: TS670 .A77 1986.

Canning Handbook on Electroplating, Published by W. Canning & Co. Ltd., 21st edition, 1970.
    This excellent book describes electroplating with many different metals and chemicals. The Canning company builds electroplating equipment. The book was written for the operator of an electroplating mill. It would be of interest to anybody who is serious about electroplating. Library of Congress number: TS670 .A1 C3.

GENERAL REFERENCES

Boylestad, Robert L., *Introductory Circuit Analysis* (10th edition; Prentice Hall College Div., 2002).

This excellent textbook is written for first and second year electrical engineering technology students. It covers basic topics on DC and AC electrical circuits. The reader is taken from the basics of Ohm's law to the complexities of three-phase phasor calculations. Library of Congress number: TK454 .B68 2003.

Fink, Donald G. and Beaty, H. Wayne (ed.), *Standard Handbook for Electrical Engineers* (14th edition; New York: Mcgraw-Hill Book Company, 1999). Purchase from McGraw-Hill Book Company, 1221 Avenue of the Americas, New York, NY 10020.

This is one of the master reference books for electrical engineers. It attempts to cover all of topics of interest to electrical engineers and does a good job of it. However, it is expensive. Library of Congress number: TK151 .S8.

Fink, Donald G. and Christiansen, Donald (ed.), *Electronics Engineers' Handbook* (3rd edition, New York: McGraw Hill Book Company, 1989). Purchase from McGraw-Hill Book Company, 1221 Avenue of the Americas, New York, NY 10020.

This is one of the master reference books for electronic engineers. It attempts to cover all the topics of interest to electronic engineers. It is expensive. Because the world of electronics is changing rapidly, it is difficult for any book to keep up. This good book should be used alongside of up-to-date product information available from electronics manufacturers. Library of Congress number: TK7825 .E34 1989.

Gibilisco, Stan, *The Illustrated Dictionary of Electronics* (8th edition; Blue Summit, PA: TAB Books, 1994). Purchase from TAB Books, Division of McGraw-Hill, Inc., Blue Ridge Summit, PA 17294-0850.

This very useful book defines electrical terms and describes many electrical devices and processes. Library of Congress number: TK7804 .T87 1994.

Palmquist, Roland E., *Guide to the 1993 National Electrical Code* (New York: Macmillan Publishing, 1993). Purchase from Macmillan Publishing Company, 866 Third Avenue, New York, NY 10022.

The National Electrical Code is a collection of suggested standards for the safe construction of electrical systems in U.S. industrial, commercial, and residential locations. The National Electrical Code can be purchased and used by itself for a little less money than a Palmquist's guide book. However, Palmquist's guide makes the code easier to understand and follow. It is worth its price. Library of Congress number: TK260 P33 1993.

Tubbs, Stephen P., *Experiments for Electrical Machines, Drives, and Power Systems* (3rd edition, copyright 1997, reprinted 2002).

My book was formerly published by Prentice-Hall, Inc, 1991 copyright. Now I publish it. Purchase it from Amazon.com and other book sellers.

The book is a laboratory manual designed for use in undergraduate university electrical technology programs. It contains 22 experiments on motors, alternators, generators, and motor control circuitry. Dewey Call Number: 621 31 T81.

Wildi, Theodore, *Electrical Machines, Drives, and Power Systems* (5th edition; Englewood Cliffs, NJ: Prentice Hall, 2002). Purchase from Prentice-Hall, 113 Sylvan Avenue, Englewood Cliffs, NJ 07632.

This excellent textbook is written for sophomore electrical engineering technology students. It covers topics on motors, alternators, generators, switchgear, power electronics, and power station equipment. It goes into calculations on all of these topics. Library of Congress number: TK2182.W53 2002.

Thomas Register of American Manufacturers 2001, Thomas Publishing, One Penn Plaza, New York, NY 10119 1-212-695-0500

This is a collection of 31 volumes that lists manufacturers of most products made in the U.S. and Canada. For the electrical experimenter the Thomas Register offers a vast information resource. Most of the topics covered in this book were partially researched by information from the Thomas Register. It is indexed by product and by name. It provides manufacturers' names, addresses, and telephone numbers. The manufacturers can then be contacted and requested to send whatever printed advertising information they have on the products you are interested in. The Thomas register can be found in most larger libraries.

Many of these books are expensive. The handbooks by Fink are over $150.00 each. However, there is a less expensive way. It may be worthwhile to search for older used editions. With the exception of the electronics area, electrical knowledge is not changing rapidly. A ten or twenty year old electrical book may be satisfactory for your needs.

SOCIETIES, GROUPS OF INTEREST, AND INTERNET SITES

American Solar Energy Society
2400 Central Avenue
Suite G-1
Boulder, CO 80301
1-303-443-3130
http://www.ases.org
> An association dedicated to advancing the use of solar energy.

American Wind Energy Association
122 C Street NW
Suite 380
Washington, DC 20001.
1-202-383-2500
http://www.avea.org
> An association that represents wind energy as a technology that is economically and technically viable today.

ENERGY EFFICIENCY AND RENEWABLE ENERGY CLEARINGHOUSE
http://www.ecomall.com/activism/renew.htm
> A long list of energy organizations. This is a good starting point for an internet search for an energy topic.

SOURCE FOR RENEWABLE ENERGY
http://www.mtt.com/theSource/renewableEnergy/index.html
> A browsable directory of over 5,500 renewable energy related businesses.

Institute of Electrical and Electronic Engineers
3 Park Avenue
17th Floor
New York, NY 10016-5997
1-800-678-4333
http://www.ieee.org
> An international organization of electrical and electronic engineers. It has subsections dealing with all application areas of electrical engineering from electrical power to micro-electronics and computers.

IEE HOME PAGE
http://www.iee.org.uk
    A starting point for exploring the many groups, societies and activities of the Institute of Electrical Engineers, the British electrical engineering society.

USENET NEWS GROUPS

alt.energy.homepower

alt.energy.renewable

alt.solar.photovoltaic

alt.solar.thermal

alt.engineering.electrical

INSTRUMENTATION

The DMM or Digital MultiMeter is the most useful instrument for the electrical experimenter. Prices have come down on them so that I no longer consider buying analog multimeters. Although, I still sometimes use one analog multimeter that I bought some years ago. The capabilities of DMMs go up with price. At the high end DMMs will connect into a personal computer for data recording, measurement of capacitance, measurement of signal frequency, and measurement of AC and DC current. All DMMs will measure AC and DC volts and ohms. Many will operate over a wide AC frequency range. You will need at least one DMM or its analog equivalent to do the experiments of this book.

Clamp-on AC ammeters are useful to those working with 50-60 Hz power circuits, such as those powering electric motors or heaters. They allow the experimenter or repairman to quickly and safely determine the AC current going through a conductor without breaking the circuit to connect in a series ammeter. I don't use one often, but it is real time saver when I do.

Wattmeters are used to determine power in AC circuits. The wattmeters I used were the analog type and were rated to operate at frequencies around 60 Hz. They worked well measuring the power into or from 60 Hz AC electric motors or alternators. The big drawback to wattmeters is that they are expensive.

Utility electric service watthour meters may be useful to some. They are designed to measure 60 Hz AC energy. Each electric utility service entrance has one. There are literally millions in the U.S., many of them recently removed from service. A surplus watthour meter could be used to determine the AC electric energy consumed or produced. A watthour meter in combination with a stop watch could be used as a watt meter.

An oscilloscope can be very useful to the experimenter. The oscilloscope allows the experimenter to see the shape of an AC voltage waveform for a wide range of AC frequencies. It should be at least a dual-trace type. The dual-trace type allows the experimenter to see the angular separation between synchronized AC waveforms. Using a current shunt and a dual-trace oscilloscope the experimenter can take measurements that will allow him to determine electrical power into or out of electrical devices. There are two drawbacks to oscilloscopes. First, they are hard to use. Ask somebody who has used one. It is sometimes difficult to get a good trace (waveform line) on the screen. Second, they are expensive. Even with the drawbacks, it can be worthwhile for the amateur to get one, provided the price is reasonable.

NEW, USED, AND SURPLUS EQUIPMENT DEALERS

Telephone or write for free catalogs from the following.

ALL ELECTRONICS CORP.
P.O. Box 567
Van Nuys, CA 91408
1-888-826-5432
http://www.allcorp.com/
Sells new and surplus electronic parts and supplies by mail order.

Digit-Key Corp.
701 Brooks Avenue South
P.O. 677
Thief River Falls, MN 56701-0677
1-800-344-4539
http://digikey.com/
Sells new electronic parts and supplies by mail order. Digit-Key's catalog has a large selection.

Edmunds Scientific Company
60 Pearce Ave.
Tonawanda, NY 14150-6711
1-800-728-6999
http://www.scientificsonline.com/
Sells scientific equipment to educators, students, and inventors by mail order. Covers most areas of science.

Herbach and Rademan
353 Crider Avenue
Moorestown, NJ 08057
1-800-848-8001
http://www.herbach.com
Sells new electronic parts, instrumentation, kits, and equipment by mail order.

Hosfelt Electronics Inc.
2700 Sunset Blvd.
Stuebenville, OH 43952-1158
1-888-264-6464
http://www.hosfelt.com/
Sells new electronic parts, instrumentation, kits, and equipment by mail order.

Jameco Electronic Components
1355 Shoreway Rd.
Belmont, CA 94002
1-800-831-4242
http://www.jameco.com/
Sells surplus and new electronic parts, computer parts, and equipment by mail order.

Newark Electronics
1-800-463-9275
http://www.newark.com/
Newark Electronics sells new electronic and electrical parts. They publish a large catalog on paper and CD. Even if you don't buy from them, their catalog is worth having as a reference. The address and phone number given above is for their administrative office. They have branch offices in many U.S. and Canadian cities, check your phone book.

RadioShack Corporation
Fort Worth, TX 76102
http://www.radioshack.com
Sells new electronics parts and instrumentation from stores all over the country. They will also do mail order sales via their on-line-catalog. Many of the projects in this book used Radio Shack parts and meters.

W. W. Grainger Inc.
1-888-361-8649
http://www.grainger.com/
Grainger sells equipment to industrial and commercial customers. Their products range from electric motors and oscilloscopes to office furniture. They publish large catalogs on paper and CD that are worth having as references, even if you don't buy from them. They have many branch offices all over the U.S., check your phone book.

USEFUL INFORMATION

RESISTOR COLOR CODES

NUMBERS AND MULTIPLIERS						TOLERANCE	
BLACK	0	1	GREEN	5	100000	NO COLOR	20%
BROWN	1	10	BLUE	6	1000000	SILVER	10%
RED	2	100	VIOLET	7	10000000	GOLD	5%
ORANGE	3	1000	GRAY	8	100000000		
YELLOW	4	10000	WHITE	9	1000000000		
			SILVER		0.01		
			GOLD		0.1		

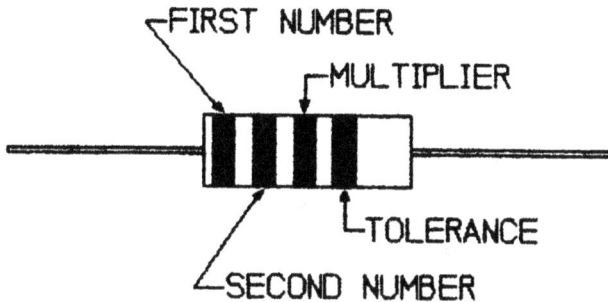

Example:
FIRST NUMBER = YELLOW, SECOND NUMBER = VIOLET,
MULTIPLIER = BROWN, TOLERANCE = SILVER
Resistor value = 47 x 10 = 470 ohms ±10%

****

AMPACITIES OF WIRE vary with the situation the wire is used in. The principle variables that affect a wire's ampacity are the duration of current, the effectiveness of cooling, and the wire's insulation heat capability.

The following chart gives the approximate continuous current AMPACITIES OF PLASTIC INSULATED COPPER WIRE:

AWG #	AMPS	AWG #	AMPS	AWG #	AMPS	AWG #	AMPS
4	85	10	30	16	10	22	3
6	65	12	20	18	7	24	2
8	45	14	15	20	5	25	1

****

****

AVERAGE DIELECTRIC STRENGTHS

MATERIAL	VOLTS/MIL	MATERIAL	VOLTS/MIL
AIR	75	RUBBER	700
CERAMIC	75	PAPER (PARAFFINED)	1300
PORCELAIN	200	GLASS	3000
TRANSFORMER OIL	400	MICA	5000
BAKELITE	700		

****

UNIT CONVERSIONS

°F	°C	°F	°C	°F	°C	°F	°C	°F	°C
30	-1.1	70	21.1	110	43.4	150	65.5	190	87.6
35	1.7	75	23.9	115	46.1	155	68.3	195	90.4
40	4.4	80	26.6	120	48.9	160	71	200	93.2
45	7.2	85	29.4	125	51.6	165	73.8	205	96
50	10	90	32.2	130	54.4	170	76.5	210	98.8
55	12.8	95	35	135	57.1	175	79.3	215	101.6
60	15.5	100	37.8	140	60	180	82.1	220	104.4
65	18.3	105	40.5	145	62.7	185	85	225	107.2

1 Inch = 2.54 Centimeters, 1 Meter = 100 Centimeters, 1 Kilometer = 1,000 Meters, 1 Mile = 1.609 Kilometers

1 Pound = .454 Kilograms(force), 1 Kilogram = 1,000 Grams

1 BTU = 1,054 Joules, 1 Kilowatthour = 3,600,000 Joules, 1 Foot-pound(force) = 1.3558 Joules

1 Horsepower = .746 Kilowatts, 1 Kilowatt = 1,000 Watts

1 Gallon = 231 Cubic inches, 1 Liter = 1.0567 Quarts(U.S. liquid)

1 Second = 1,000 Milliseconds

1 Hertz (Hz) = 1 Cycle per second

Milli = times .001

Centi = times .01

Kilo = times 1,000

Mega = times 1,000,000

www.ingramcontent.com/pod-product-compliance
Lightning Source LLC
Chambersburg PA
CBHW080557220326
41599CB00032B/6520